AF754323

Africa, Tropical Timber, Turfs, and Trade

Africa, Tropical Timber, Turfs, and Trade

Geographic Perspectives on Ghana's Timber Industry and Development

J. Henry Owusu

LEXINGTON BOOKS
Lanham • Boulder • New York • Toronto • Plymouth, UK

Published by Lexington Books
A wholly owned subsidiary of The Rowman & Littlefield Publishing Group, Inc.
4501 Forbes Boulevard, Suite 200, Lanham, Maryland 20706
www.rowman.com

10 Thornbury Road, Plymouth PL6 7PP, United Kingdom

Copyright © 2012 by Lexington Books

All rights reserved. No part of this book may be reproduced in any form or by any electronic or mechanical means, including information storage and retrieval systems, without written permission from the publisher, except by a reviewer who may quote passages in a review.

British Library Cataloguing in Publication Information Available

Library of Congress Cataloging-in-Publication Data

Owusu, J. Henry (John Henry), 1953–
Africa, tropical timber, turfs and trade : geographic perspectives on Ghana's timber industry and development / J. Henry Owusu.
p. cm.
Includes bibliographical references and index.
ISBN 978-0-7391-7401-2 (cloth : alk. paper) — ISBN 978-0-7391-7402-9 (electronic)
1. Timber—Economic aspects—Ghana. 2. Lumber trade—Ghana. 3. Ghana—Economic policy. 4. Ghana—Economic conditions. 5. Neoliberalism—Ghana. 6. Structural adjustment (Economic policy)—Ghana. I. Title.
HD9767.G42O98 2012
338.1749809667—dc23 2012018318

™ The paper used in this publication meets the minimum requirements of American National Standard for Information Sciences—Permanence of Paper for Printed Library Materials, ANSI/NISO Z39.48-1992.

Printed in the United States of America

This book is dedicated to my late parents,
J. B. K. and Docia Owusu,
and to my family,
Marilyn, Leon, and Leslie Owusu

Contents

List of Figures and Tables

FIGURES

TABLES

List of Acronyms

AAF-SAP	African Alternative Framework to Structural Adjustment Programmes for Socio-Economic Recovery and Transformation
AGOA	African Growth and Opportunity Act
ATP	African Timber and Plywood Limited
AAC	Annual Allowable Cut
ATL	Atwima Timbers Limited
BOP	Balance of Payments
CIDA	Canadian International Development Agency
CSM	Chainsaw Milling
CPP	Convention People's Party
CEPS	Customs, Exercise and Preventive Service
DANIDA	Danish International Development Agency of Denmark
DFO	District Forestry Officer
ERP	Economic Recovery Program
EC	European Community
EEC	European Economic Commission
EU	European Union
ECGD	Export Credit Guarantee Department
FAO	Food and Agriculture Organization
FWP	Forest and Wildlife Policy
FDMP	Forest Development Master Plan
FPIB	Forest Products Inspection Bureau
FRP	Forestry Research Program
FRMP	Forest Resource Management Project
FERN	Forests and European Union Resource Network
GES	Ghana Educational Service
GEPC	Ghana Export Promotion Council
GFPA	Ghana Furniture Producers Association
GTA	Ghana Timber Association

GTMB	Ghana Timber Marketing Board
TIMBOD	Ghana Timber Marketing Board (Restructured)
GTMO	Ghana Timber Millers' Organization
GIC	Ghana Investments Center
GWA	Gliksten West African Limited
GOG	Government of Ghana
GDP	Gross Domestic Product
HF	Human Factor
ISSER	Institute of Statistical, Social and Economic Research
IDA	International Development Association
IFI	International Financial Institution
IMF	International Monetary Fund
ITTO	International Tropical Timber Organization
LPA	Lagos Plan of Action
MLF	Ministry of Lands and Forestry
MIGA	Multilateral Investment Guarantee Agency
NDC	National Democratic Congress
NRC	National Redemption Council
NPP	New Patriotic Party
NGO	Non-Government Organization
NPART	Non-Performing Assets Recovery Trust
OAU	Organization of African Unity
ODA	Overseas Development Administration
PNP	People's National Party
PNDC	Provisional National Defence Council
SECAL	Sector Adjustment Loan
SRA	Social Responsibility Agreement
SAP	Structural Adjustment Program
SSA	Sub-Saharan Africa
SMC	Supreme Military Council
SECO	Switzerland's State Secretariat for Economy
TEDB	Timber Export Development Board
TIDD	Timber Industry Development Division
TRMA	Timber Resources Management Act
TREC	Timber Rights Evaluation Committee
TUC	Timber Utilization Contract
TED	Trade and Environment Database
TA	Traditional Authority
UN/ECA	United Nations Economic Commission for Africa
UNIDO	United Nations Industrial Development Organization
UN	United Nations Organization
WRI	World Resources Institute
WTO	World Trade Organization
WW I	First World War
WW II	Second World War

Preface

In a chapter discussing Africa's natural resources, Ghana's first post-independence president, Dr. Kwame Nkrumah, in his book *Neo-Colonialism: The Last Stage of Imperialism*, noted that "Africa is a paradox which illustrates and highlights neo-colonialism. Her earth is rich, yet the products that come from above and below the soil continue to enrich, not Africans predominantly, but groups and individuals who operate to Africa's impoverishment." Sadly, in Africa's tropical rain forest zone—from Côte d'Ivoire and Ghana in West Africa, to Cameroon and Gabon in Central Africa—this has been the case of Africa's tropical timber. Various groups and individuals—international and domestic—have created turfs to control the timber trade to enrich themselves to the impoverishment of Africa, particularly the forest communities.

It was in such a community, at Goaso, that I began my primary education in Ghana, then the Gold Coast, a British colony. Goaso, a small town in the Brong-Ahafo Region, nestled in Ghana's tropical timber-rich rain forest was a few miles from Mim, a key town in that nation's timber export process, established in the late 1950s by Swiss entrepreneurs; a town founded as a hub for assembling tropical timber for preparation for export to Europe; a town known for timber processing and furniture manufacturing. Currently, major wood processing companies here include Scanstyle-Mim Ltd., which operates a furniture fabrication company, and AYUM Timber products, Ltd., which produces plywood, veneer, and lumber, and owned by one of the largest manufacturers and exporters of wood based products in the West African sub region, the Naja David Group of companies, affiliated to Amerind Pty Limited, Australia; Amerind Limited, UK; and Timb-Mak Export GmbH in Germany. Mim, however, is a paradox; Mim, to me, spells abject poverty in the face of the immense wealth generated by these companies. This is because, as observed by Project MIM, a non-profit organization, and social enterprise focused on improving lives in this rural community, "despite the presence of the timber industry, conditions in Mim are severe and life is challenging for nearly everyone. The population in the town center is dense and untreated sewage trickles from family homes onto unpaved streets. Gutters and sewer pipes simply don't exist in most neighborhoods. The resultant spread of disease and sickness due to inadequate sanitation also plays a large part on the local economy. Without gutters, getting around the muddy streets during the rainy season is simply impossible. Constant downpours push torrents of rainwater and raw sewage through streets, carving deep, impassible ditches for the remainder of the year. These ditches not only make fetching water or getting to school difficult, the zig-zagging crevasses prevent trucks from entering neighborhoods, which inhibits commerce as well as increases physical isolation and impedes growth." At the local level, the forest communities are largely marginalized and deprived of the wealth and bounty of the forest.

I continued my primary education as well as my secondary and undergraduate education in Kumasi, Ghana's second largest city and the largest and most important urban center in the tropical high forest or closed forest zone from which the country's variety of timber is extracted. It is also home to the biggest agglomeration of sawmills as well as the biggest conglomeration of informal wood processors and lumber traders in the country. Historically, it was the capital of the Ashanti Kingdom which was colonized by the British in 1902 and thus became the most important politico-administrative center and military outpost in the interior, next to the colonial headquarters, Accra, on the coast, and became the hub of the colony's transportation and communication network, the northern apex of the railroad network, geared at the extraction and export of natural resources including timber, and the most important center of commerce in the interior. My residence and education in Kumasi thus experientially exposed me to the spatial structure of the formal and informal aspects of the wood processing industry in Ghana—from its colonial roots to contemporary times, and by extension, in comparable African countries.

My graduate education in the United States—in Geography and Regional Planning at Indiana University of Pennsylvania; in Public Policy Evaluation and Analysis at the Graduate School of Public and International Affairs University of Pittsburgh; in Planning for Developing Countries at Portland State University, and in Third World Regional Development at the Geography Department, The University of Iowa—all prepared and equipped me to examine the significance of Mim and Kumasi with respect to the timber industry; to look back and take a closer look at Africa's underdevelopment and persistent poverty despite her abundance of natural resources—the paradox that Nkrumah discussed in his book on neo-colonialism. It then became apparent that for Ghana, no one had actually documented the paradox with respect to the timber industry and its role in that nation's development spanning the colonial era through the neo-liberal (essentially, *neo-colonial*) era in a comprehensive manner from a development geographer's perspective. It is the need to fill this gap in the development discourse and the lessons that could be learned by African development practitioners and students as well as people interested in tropical timber and trade in comparable African and other developing nations that inspired me to write this book.

Researching and writing this book has been possible with the advice, support, assistance, and encouragement of a number of individuals to all of whom I am very grateful. There are however, a few who deserve special mention. First, my profound appreciation and gratitude go to my mentors, Professor Emeritus Michael McNulty, formerly of The University of Iowa and Professor Abdi Samatar of University of Minnesota and Vice President of the African Studies Association, who significantly helped shape my intellectual perspectives; Professor Emeritus George Benneh, former Vice-Chancellor of The University of Ghana, Legon; Ghana's former Minister of Lands, Natural Resources, Fuel and Power; and also former Minister of Finance and Economic Planning; and Professor Emeritus S. Tetteh-Addo of the Department of Geography and Resource Development, also of The University of Ghana, Legon, for their assistance and support during my fieldwork in Ghana. I am also indebted to Messrs. Agyei-Yeboah, former Deputy Minister of Lands, Forestry and Mines, Ghana, the late Maxwell Owusu Jr. of MOW Timbers, Ghana, and Kofi Asante-Okyere of The Makers Ltd., all of whom gave me a very deep insight into Ghana's timber industry. I am very grateful to them for the assistance they gave me in diverse ways in getting the information I wanted.

In the preparation of this manuscript, I have also benefitted from the Graduate College, the College of Social and Behavioral Sciences, and my academic department, the Department of Geography, University of Northern Iowa, Cedar Falls, Iowa. This project was supported in part by the Graduate College via a 2004 UNI Faculty 4-Week Summer Fellowship and a Fall, 2007–2008 Professional Development Assignment, and for these I am very grateful. I appreciate and cherish the support and encouragement I received from Professor Philip Mauceri, Dean of the College of Social and Behavioral Sciences, and several of my colleagues. Special thanks go to my mentor, Professor David May, who has been of immense help to me in the department in diverse ways and to Ms. Abby Bachman, the department's student Office Assistant who worked very hard helping me develop all my graphs, charts and maps to meet the publisher's specifications.

Figure 2.1 was used by permission of the cartographic section of the United Nations Organization, and figure 3.4 was also used by permission of the University of Texas Libraries, The University of Texas at Austin, and I am very grateful to them. My deep appreciation goes to the Center for International Forestry Research (CIFOR) for inviting me to participate in the *Workshop on Forest Governance and Decentralization in Africa*, held in Durban, South Africa in April 2008. This initiative in support of the United Nations Forum on Forests gave me much more insight and input for the manuscript, for which I am grateful. Thanks also to my publishers, Lexington Books (Rowman and Littlefield Publishing Group), especially Ms. Melissa Wilks, Associate Editor and Ms. Alison Northridge, Assistant Editor, Acquisitions in the Political Science, Religion and Geography Division, and Emily Natsios, Assistant Editor, Production. I gratefully acknowledge the anonymous reviewers whose comments and suggestions greatly helped to improve the quality of the manuscript.

I would also like to give special thanks to Professor Emeritus Ruth Shirey, formerly of Indiana University of Pennsylvania and my brother, Martin Owusu of Silver Spring, MD, who facilitated my graduate studies in the U.S. My deep gratitude also go posthumously to my dear late parents, J. B. K. and Docia Owusu, whose sacrifice, support, guidance, and encouragement largely made me who I am; and to my siblings, Enoch, Lord C. J., Dora, Miriam, Regina, Sammy, Nana Aduwaa, and Kwasi Ketewa, and also my great friend, Dr. Justin Bediako-Poku, who have all been a great source of support and encouragement. Finally, I would like to express my utmost gratitude to my dear wife, Marilyn, and to our two boys, Leon and Leslie for their love, patience, understanding, sacrifice, support, and encouragement.

Cedar Falls, Iowa, U.S.A.
January 25, 2012

Introduction

This book examines the issue of structural transformation in Sub-Saharan Africa—the positive changes or progress made in the region's economies, resulting from domestic and international policies, such as modernization and neo-liberal structural adjustment programs—aimed at correcting the systemic economic and social imbalances that obstruct development in the region. This objective is tied to the perspective that, the root of Africa's persistent problem of underdevelopment is the very structure of its political economy. Several academics and practitioners of Third World development have written on the issue to the extent that another book on the subject is likely to be perceived as a mere rehash of the same old analyses and conclusions. This book takes a fresh and different approach in the sense that it is a comprehensive geographical study, primarily from a structuralist perspective within that discipline. For the reader without a formal background in the discipline, Geography, in a nutshell, studies the distribution of both physical and human phenomena on the surface of the earth, the interactions within and among the phenomena, and the effects of the interactions, which could be examined at various scales of abstraction—local, national, regional or global, over time. The structuralist perspective, which examines development issues through social relations couched in a historical context and political economy, thus provides us with a framework whereby we are able to look at structural transformation in Africa by:

- recognizing the complex interactions and interrelations among the Western capitalist nations, particularly Western Europe, and Africa;
- couching the changes in a historical context spanning developments in Europe that facilitated the eventual incorporation of pre-capitalist Africa into the international capitalist system to show how the structural elements, such as the formal and informal sectors of African economies evolved;

- analyzing the locus of power both at the inter-national and intra-national levels, especially with respect to indigenous control of national destiny, or development as reflected in the extent of political and economic independence; dependency, disparities; and related processes like tensions, conflicts, crises; and
- considering the very nature of the process whereby the changes—economic growth and development—are achieved, their associated beneficiaries, winners and losers, as well as such implications like the environmental effects of the change.

Finally, the structuralist perspective will enable us to better understand the nature of inter-national and intra-national relations among African and Western societies generated by the processes and forces of production such as the:

- direct relationships as between African and Western capitalists, producers, consumers, and traders in the form of formal and informal domestic, regional, and international markets;
- formal and informal groupings into social classes or interest groups;
- formal organizations like corporations and government agencies; and
- the related economic values, motivations, and ideologies.

As a comprehensive geographical study, the issues raised and discussed in this book are couched particularly in economic geography and regional development, as the book addresses the effects of national and international development policies on the process of industrialization and economic development in Sub-Saharan Africa (SSA), with a particular focus on Ghana. Within this broad policy framework of structural change or transformation, Ghana's wood processing industry, as an integrative sector of the economy, is utilized to analyze the SSA's development as it relates to:

1. Spatial (geographical) structure involving the location and distributional characteristics of firms, and their qualitative and quantitative characteristics regarding material inputs, the environment, product outputs, labor, markets, capital, machinery, and equipment;
2. Spatial and sectoral linkages involving inputs, outputs, inter- and intra-sectoral exchanges, and exports; and
3. Organizational structures and processes involving related public and private institutional structures, their relations with firms, and firm ownership structures.

In this regard, the book documents the structural and organizational changes or transformations that have occurred in Ghana's timber industry, which typifies the nation's structural characteristics, from the Colonial era up to the beginning of the twenty-first century—specifically around 2005–2006, the tail end of the J. J. Rawlings' National Democratic Congress (NDC) administration, into the beginning of the J. A. Kufour's New Patriotic Party (NPP) administration. Thus it covers the period when Ghana, then the Gold Coast, was incorporated into the global capitalist system as a British territory, through the post-independence period ending in the neo-liberal dispensation, reflecting the effects of the new roles played by powerful international bodies like the International Monetary Fund (IMF), The World Bank, and the World Trade Organization as agents of economic change. Particular emphasis is placed on the nature and effects of the country's economic recovery (structural adjustment) program, initiated in 1983, under the auspices of the World Bank and the IMF.

"Tropical Timber, Turfs, and Trade" in the book's title is used to capture the discussion of the role played by a variety of *Tropical Timber,* which are an important renewable natural resource with long gestation periods, whose extraction generates significant environmental issues like deforestation. As a resource with certain unique industrial, aesthetic, and other attractive qualities, the production and consumption of tropical timber are carried out in various *Turfs*, the operational milieux expressed in various geographic scales, from the domestic, regional to the international. There is first, the natural turf, namely the geographic or ecological regions where the tropical timber grows, and which can be sustained or compromised by human actions or inactions; and secondly, the socially created turfs, with physical and politico-economic dimensions which may be formal or informal. Essentially the formal turfs are created to enhance the operations of international capital, and in this respect, the national government's turf falls within its ambit, while informal turfs develop in reaction to them, particularly and directly to the latter. As for turfs associated with traditional authorities, they tend to straddle both the formal and informal ambits. Like any socially created turf, these turfs are formally or informally actively controlled, preserved, protected, and expanded, if expedient. For example as we shall see in our discussion, for purposes of control to meet the imperatives of capitalism, such as profit and power, international capital, through certain formal or informal institutions, sought to actively create, preserve, and in fact protected, and expanded its production and consumption turfs through a variety of policies, which arguably, even tended to be structurally violent in certain cases. Again, as with any socially created turf, under certain circumstances, they may be formally or informally actively challenged, contested, or encroached upon by other groups, as each group seeks to advance its own interests. These groups may be formal such as national governments, domestic or international capital, interest groups like environmentalists and the like, or informal like peasant farmers and illegal chainsaw loggers and millers. The dynamics of the production and consumption of the tropical timber which are carried out in the various turfs are linked with certain important processes, interactions, and effects, which are subsumed under domestic (formal and informal), and international *Trade*, which involve issues like organization, imports, exports, and related policies.

Our discussion, captured as part of the title *Tropical Timber, Turfs, and Trade*, is therefore couched in the role that geography and its conceptual and theoretical contributions to the structural transformation debate play in relation to: (a) the unequal relationship between the advanced and developing economies of the world, and (b) how that affects economic development and trade behavior of developing economies. Consequently, the discussions in this book are part of the exploration of the dynamics of global economic restructuring and adjustment as a way of further integrating developing economies into the global capitalist economy—a contribution to the current debate in Development Geography/Development Studies regarding the dynamics of international and domestic capital and their implications for Third World economies at three levels: at the sectoral level, it ties together empirical studies of the developmental role of formal and informal wood processing industries in Ghana during the colonial and post-colonial periods and their implications for comparable African countries. At the national level, it addresses the question of whether or not there has been a deviation from the colonial patterns of economic development, after Ghana gained its political independence in 1957, and identifies, analyzes, and discusses the associated exogenous and indigenous factors mitigating or compounding the colonial influence; and at the regional level, it incorporates an examination of the effectiveness of trade in the Economic Community of West African States.

The discussions in the book integrate most of the author's research and associated publications from 1993 to 2006 on the impact of colonial and post-colonial international and national development policies on industrialization and economic development in Ghana in particular, and Sub-Saharan Africa in general. It analyzes African development /underdevelopment in the colonial and post-colonial periods, using Ghana's timber industry as the primary focus. This focus is based on the fact that the timber industry, particularly wood processing sector is characterized by the typical spatial structural issues associated with underdevelopment such as excessive external dependence for capital, markets, etc.; minimal internal functional interconnections or linkages; and lop-sided or inequitable regional development. In addition, wood-processing serves as an integrative sector of the economy linking the natural resources extraction, and formal and informal manufacturing production activities. This link has several important geographical implications involving, *inter alia*, the use or abuse of the environment, internal and external trade, as well as public and private sector involvement. Consequently, the forestry sector serves as an excellent sector for examining structural transformation in Africa. The wood-processing sector also meets the objective of documenting the post-colonial changes in the structure and organization of the industry and their implications for development.

This book is divided into nine chapters. Chapter 1, Africa, Turfs, and Trade: Theoretical and Historical Framework, provides the theoretical framework employed in the discussions in the book, and presents the various concepts and processes used to explain the issues which are raised in the book, including the related conceptualizations of space and such relevant historical processes such as colonialism. It also identifies spatial integration as a critical goal of structural transformation in Sub-Saharan Africa, given the fact that all advanced countries are largely functionally linked internally in terms of their economic activities. Chapter 2, Turf Establishment and the Development Process: Structural Foundations of Ghana's timber industry, focuses on the development of Ghana's timber industry, the role it plays in the Ghanaian economy and the nation's development process, and discusses the colonial process and the associated creation of a capitalist space or turf based on the exploitation of timber and the beginnings of the nature of the tropical timber trade internally and externally. In this regard, it concentrates on the pre-independence developments involving the changes that took place in the pre-colonial and the colonial eras, ending in 1957, when Ghana secured its independence from the British. Chapter 3, Independence, Instability, and Contested Turfs: Structural Changes in the Timber Industry, continues the discussions in chapter 2 by looking at key developments in the industry in the post-independence period through the beginning of the twenty-first century. Here, the discussion focuses on the post-independence socialist policies of the Kwame Nkrumah administration to the structural patterns that had emerged by 1983 which necessitated the adoption of neo-liberal policies at the behest of the World Bank and the International Monetary Fund, by the Jerry Rawlings administration, which had morphed from a military dictatorship. This is done with respect to how these policies affected the timber trade and the associated internal and external turfs. An important aspect of this chapter is that it provides a detailed presentation of the state of the industry in terms of its problematic structural and organizational characteristics by 1983, which made it a priority target under Ghana's structural adjustment program, dubbed "Economic Recovery Program" (ERP), as a foundation for the discussions in subsequent chapters relative to the issue of structural transformation.

Chapter 4, Turf Reclamation Strategies: Neo-Liberalism and the Timber Industry, looks at neo-liberalism, the Ghanaian economy, and the political economy of Ghana's structural adjustment program. The competing perspectives on structural adjustment relative to the Ghanaian economy are presented. Chapter 5, Turf Recapture and Consolidation for the Timber Trade: Ghana's Economic Recovery Program, discusses the adoption and implementation of Ghana's adjustment program, the ERP. It also outlines the neo-liberal policy measures relating to the timber industry, and an evaluation of the policy measures. Chapter 6, Structural Adjustment, the Timber Trade, and Turf Degradation: Political Convenience and Desperate Deforestation, discusses the link between the neo-liberal policies, the timber trade, and Ghana's unprecedented deforestation, identifying the important role played by corporate and national debt, particularly from international loans. Chapter 7, Turf Dynamics and Conflicts, the Timber Trade, and Spatial Integration, focuses on the determinants of production within the industry, including human performance, and examines structural transformation and the associated dynamics resulting from the neo-liberal policies and how those policies addressed the goal of spatial integration. In relation to these, formal-informal sector turf conflicts, corruption, and the marginalization of the forest communities are discussed. Chapter 8, Debt, Dependence, Turf Control, and Ghana's Economy, focuses on the dynamics of the corporate and national debt and dependence relative to the timber industry and the Ghanaian economy in general; and their implications for comparable African countries. The final chapter, chapter 9, Conclusions and Lessons for Twenty-First Century Africa, concludes the discussion by tying together the issues of development in Africa with the tropical timber trade, and the related dynamics of formal and informal operational turf creation and control, and identifies and distills the lessons that can be learned from Ghana's experience for the benefit of students and practitioners of development in Ghana and comparable African and other developing countries in the twenty-first century.

This book is a contribution to development studies literature at two levels: at the level of African development literature, there is a dearth of studies that have specifically examined the changes that have occurred in Ghana's timber sector since the country's independence in 1957, in a comprehensive and empirical manner. In part, this book seeks to fill this gap by focusing on the spatial structure of African economies and some of the associated dynamics, particularly with respect to Ghana. As a contribution to the Third World development literature, the book is intended to serve as an important resource in three broad areas, namely African Development Studies, Tropical Forestry, and in Economic Geography. This book is thus comprehensive in its scope and is intended to serve as an important resource and reference material or document on Ghana /Africa's tropical timber resource exploitation and trade, and their implications for development such as deforestation, the role of international forces, the informal sector, and the timber industry's changing markets. It is also intended to significantly add to the body of knowledge and improve tropical-timber related policies and practices because:

- It takes a fresh and different approach in the sense that it is a comprehensive geographical study, spanning the colonial and postcolonial eras, of a typical resource-exporting African economy, primarily from a structuralist or political economy perspective within that discipline; and
- It fills a gap that has not been addressed to date in the Third World development literature, particularly during this era of rapid globalization, by serving as an important resource in African Development Studies, Forestry, Economics, and in Economic Geography.

In sum, this book documents the structural and organizational changes or transformations that have occurred specifically in Ghana's timber industry, which typifies the nation's structural characteristics, from the Colonial era up to the beginning of the twenty-first century. However, it is intended primarily for:

1. Undergraduate, post-graduate, and scholarly levels as a supplementary reading in Economic Geography, Development Geography, Development Studies, Tropical Forestry, Economics, Resource Economics, and African Studies;
2. African development policy-makers, Third World development practitioners, and environmentalists—in other words, actors and stakeholders in the tropical timber industry and trade;
3. Tropical forestry research institutes;
4. International forest policy-makers and African government officials dealing with tropical timber and trade, and
5. Researchers and general readers interested in Africa, African development, tropical deforestation, colonialism and neo-liberalism, tropical timber trade, Formal and informal sectors in Africa and the Third World, and economic and political activities of international institutions like the World Bank, the International Monetary Fund in Africa, and the Third World.

It is for this broad audience that the subject matter of the book and discussions are geared.

Chapter One

Africa, Turfs, and Trade

Theoretical and Historical Framework

HISTORICAL BACKGROUND AND RELATED CONCEPTS

This chapter discusses the structural foundations and the associated theoretical issues and concepts that underpin contemporary African economies and how they operate. It provides the basic framework for our discussion since, as noted by Griffiths (1995, p. 1), understanding African economies and how they function must necessarily be put in the context of basic structural defects. Structures in this respect can be defined as the economic, political, and socio-cultural features or institutions in place, which facilitate, enhance, or constrain national development or progress. These may be in the form of such features as established practices, agendas, customs, and laws. These features or institutions may not in all cases be readily apparent; they may be very subtle in some of their manifestations. They may be endogenous in a number of cases or they may be exogenous or a combination thereof. Structures are therefore the underlying and enduring driving forces within society as defined by Holt-Jensen (1999, p. 227).

Structural problems or defects, in the context of Third World development, generally refer to the economic and social imbalances that obstruct development (Engberg-Pedersen et al., 1996, p. 3). In Ghana, our case study, as in other Sub-Saharan African countries, structural problems include the raw materials export-bias of the region's economies, and the associated poor functional interconnections or linkages within and among the economic sectors, which are all largely legacies of the colonial era. In acknowledging this problem, the United Nations Economic Commission for Africa (UN/ECA), for example, noted that "the structure of the African economy defines the essential features of Africa's central problem of underdevelopment" (UN/ECA, 1986, p. 1). This was based on its observation that the general post-colonial structure of the region's economies had continued to be "colonial." This prompts a preliminary definition of colonialism, which in essence, could be described as

> the extension of a nation's sovereignty over territory beyond its borders by the establishment of either settler colonies or administrative dependencies in which indigenous populations are directly ruled or displaced. Colonizing nations generally dominate the resources, labor, and markets of the colonial territory, and may also impose socio-cultural, religious and linguistic structures on the conquered population (New World Encyclopedia).

Africa's structural problems are linked to the region's colonial history because colonialism fundamentally re-ordered the economic structures within each of the colonies to integrate them into the world capitalist system. In the process, the dynamic sectors of each economy were aligned with the interests of the colonial powers (Sawyerr, 1990, p. 10). In this regard, a report produced for the Non-Government Organization (NGO), Forests and European Union Resource Network (FERN) by Forest Watch Ghana, *Forest Governance in Ghana* (2006) noted regarding a crisis characterizing the forest sector that:

> Forest policy, forest legislation, the forestry bureaucracy, and the rest of the governance superstructure all have roots in the colonial experience, and continue largely to be determined by the substructure of exploitative international economic relations that colonialism introduced to Ghana between 1850 and 1950. Post-colonial changes in this substructure are responsible, however, for the specific features of today's crisis (p. 19).

The examination of Africa's structural issues relating to the timber industry thus elicits a brief discussion of the history of capitalism with respect to Africa in terms of the creation of capitalist spaces or "turfs" through the process of colonialism and neocolonialism as capitalist relations of production evolves over time. To do this, let us examine the nature of the African space for our purposes, the nature and dynamics of capitalism and the imperative of colonialism for the creation and reproduction of capitalist spaces or territories for accumulation in Africa.

AFRICA AS A CAPITALIST SPACE: THREE INTERRELATED SPATIAL CONCEPTS

As noted by Gore (1984) and Mabogunje (1981), the African space, as an operational milieu for capitalism, can be conceptualized in three ways that are really not mutually exclusive, since any geographical space can become one or all simultaneously depending on circumstances or issues being examined. The three conceptions are Africa as: 1. *Absolute* space; 2. *Relative* space and more importantly, 3. *Relational* space.

The absolute and the relative concepts, although useful in certain analyses, are limited by the fact that they separate spatial patterns from social, political, and economic processes, which are dynamic in nature. This is because the absolute conception deals with the physical attributes of Africa as a continent—a container space; as the physical context or environment within which the African people operate with its associated opportunities and constraints. Africa in this respect is conceptualized as a continent—a distinct, physical, and imminently real or empirical entity, as defined by Holt-Jensen (1999, p. 216). The relative concept, on the other hand, deals with the location of, and the relationships between different physical areas, regions, and other phenomena in terms of distance *on* the African continent and relative to other areas or regions *external* to it, as measured using such indices as transport costs, mileage, or temporal proximity. The absolute and the relative concepts thus merely show the static spatial patterns evident on or relative to the continent, i.e., patterns frozen in time and space or snapshots of what was or currently exists.

The "relational" conceptualization of the African space is central to our discussion. In this context space, which can be examined at various scales of analysis—continental, regional, national, and intra-national—is regarded as an integral element of social interaction (Gore, 1984, p. 181). It must be noted that at the heart of our discussion is the issue of African development, which however defined involves a *process* of change. It is thus within this context of a functional link between space and process that the broad African space and its constituent spatial units such as regions, countries, and even their economic sectors are described as "turfs." This is because each component spatial unit like a country or the informal economic sector of a country is a socially defined relational space, and may have absolute and relative dimensions. As with any socially created turf, the African space and its constituent spatial units tend to be formally or informally actively controlled, preserved, protected, and expanded by the related actors, if expedient with implications for the related stakeholders. Again, as noted earlier, as with any socially created turf, the African space and its constituent spatial units tend to be formally or informally actively challenged, contested, or encroached upon by other groups. The African space, in this sense, is thus part of the broader global spatial system, in which space essentially interrelates rather than separates spatial patterns and social processes (Gore, 1984, p. 77).

Using the concept of relational space therefore enables us to factor into our discussion social relations and structural issues such as power—economic, political, military—in terms of dominance, exploitation, etc., and sociocultural institutions, etc., that facilitate, enhance, or constrain development in Africa. This is because the "African Space," from this perspective, can be regarded as malleable, shaped by the functional interrelationships considered in our discussions—a "shaped matter, acquiring specific forms and styles of disposition depending to a large extent on the prevailing mode of production" (Mabogunje, 1981, p. 52). This is where capitalism as the contemporary mode of production enters the equation, as it created its own spaces for accumulation in Africa through the historical processes of imperialism, colonialism, and lately, the reproduction of such spaces through neo-colonialism packaged as neo-liberalism. The importance of the use of the relational space concept here is because, (a) as noted by Ninsin (1989), the global economy, of which Africa is part, is dominated by the law of capitalist accumulation—the capitalist law of value, involving the immanent tendencies of the capitalist system and which determines the range of options open to individual capitalists, corporations, and to national economies; and (b) colonialism, the process which facilitated the incorporation of African economies into this world system. Consequently, they stand in a dialectical relationship to the rest of the global economy, and the effect of Africa's participation in this has been what Frank (1967) called the "development of underdevelopment." In order to appreciate the implications of the relational space thereby established, we need to look at the processes involved in the creation and reproduction of capitalist spaces or territories for accumulation in Africa and their dynamics.

IMPERIALISM, COLONIALISM, AND THE CREATION OF CAPITALIST TURFS IN AFRICA

As noted by Sawyerr (1990, p. 10), African economies and the way they operate are defined by three factors: (a) their colonial experience, (b) the international relationships that were thereby established, and (c) the relationship of both to contemporary domestic and external economic structures. To relate these to our discussion, let us look at each factor as a basis for further discussion in subsequent chapters:

The Colonial Experience

All African countries except Liberia and Ethiopia were formally colonized by European powers. Even in the case of these two exceptions, it has been argued that "despite their formal independence, the subjection of their political economy to external political and economic domination was as thorough as the formal colonies, and they can for this purpose be considered in the same light" (Sawyerr, 1990, p. 10). To the extent that colonialism involves the forceful extension of various European nations' political, cultural, and economic authority over pre-colonial Africa by seizing the latter's territories, or establishing political and economic dominance over those territories, colonialism is essentially, imperialistic. As an expression of imperialism, colonialism is defined as "the establishment of domination of a geographically extended political unit, most often inhabited by people of a different race and culture, where this domination is political and economic and the colony exists subordinated to and dependent on the mother country" (Wolpe, 1975, p. 231). It was this process whereby European capitalist nations extended their sovereignty over various parts of Africa that served as the mechanism by which Africa was formally incorporated into the global capitalist system (Slater, 1975) and transforming it into a capitalist space. It must be noted here that the incorporation did not mean total absorption, as part of the colonies continued to be pre-capitalist spaces. As pointed out by Dicken (1988, pp. 95–96), the pursuit of profit constitutes the driving force in a capitalist economy—"the 'laws of motion' of capitalism derive primarily from the drive to increase profits and to accumulate capital through the increasing appropriation of surplus value from the process of production." The aggressive imperialistic moves by European powers—especially Britain, France, and Belgium—dubbed the Scramble for Africa between 1880 and 1914 to partition and colonize the region were thus a logical expression of the dynamics of capitalism.

What then stimulated the imperialistic moves by European powers to partition and colonize the region? In other words, why did such capitalist turfs become necessary for Europe?

Essentially, as illustrated in figure 1.1, given the cyclic nature of capitalist relations of production over time—manifested by diminishing profits and the need for land, labor, and resources—Africa had certain relatively unique geographic and historical attributes that made that region attractive to Europe. Given those features, various places on the continent offered a variety of opportunities not only for capital (e.g., exploitation of raw materials, cheap labor, and markets) with respect to the profit imperative, but also for sociocultural transformation, such as the introduction of Western culture and its associated changes in the traditional African structures.

Potter et al. (1999, p. 33) underscore the reasons for the scramble for Africa, by identifying two associated processes—external and internal—noting that "not all the impetus came from outside the continent, although much of it did." As illustrated in figure 1.1, they acknowledge the roles

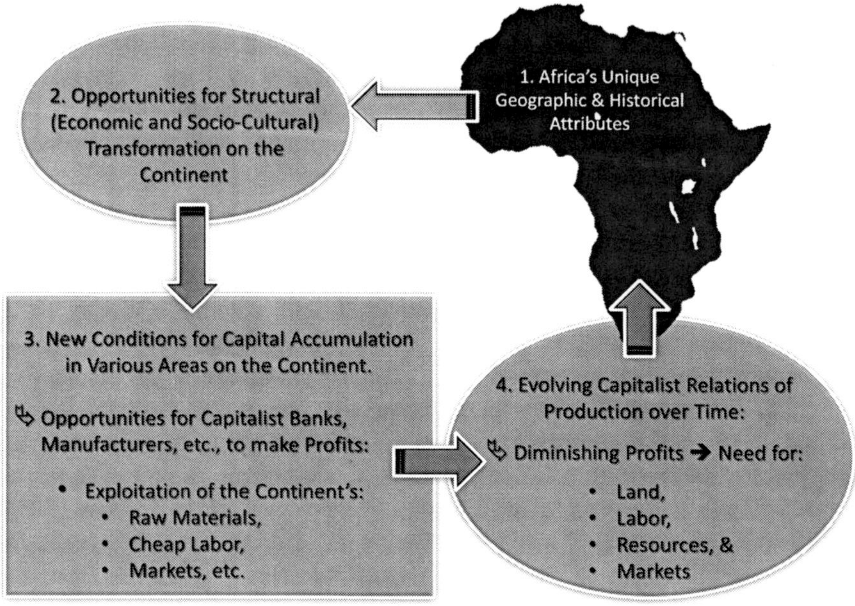

Figure 1.1. Cyclic Processes in Africa's Socio-Spatial Evolution

of imperialism and capitalism and note that some of the processes were chronologically and spatially specific. As regards the external processes, they point out that:

1. Between the first and second industrial revolutions (i.e., the shift from coal and iron to oil, electricity, and steel), Europe went through a deep recession. *As the rates of profit fell, European firms began to seek new material sources, new markets, and new investment opportunities on an extensive scale, partly to forestall other European rivals.*
2. The last quarter of the nineteenth century saw the newly united countries of Italy and Germany, using colonialism to sidestep internal tensions. *As Africa was the largest uncolonized area, it became the site of a national scramble for territory and prestige,* with France seeking compensation for defeat by Germany within Europe.
3. These processes were facilitated by a *technological revolution,* particularly in transport, where steamships, railways, and telegraphic links *accelerated both decision making and physical advances into Africa.* New armaments, such as machine guns, facilitated this process by smaller and smaller European forces.
4. Once acquired, *many colonies were also seen as healthy places for the surplus European population* that technological advances in medicine and hygiene were beginning to produce.

Colonialism, therefore, was the mode for creating capitalist relational spaces or "turfs" in the region (e.g., Colonies A_1 and A_2 for Metropolitan Economy A, as shown in figure 1.2). The partitioning of Africa into colonies, and the establishment of these turfs, as exclusive spheres of influence in Africa provided the Europeans with relational spaces that enabled them to dominate and regulate the resources, labor, and markets in those spaces in consistency with the goals of increased profits and capital accumulation associated with the dynamics of capitalism. These spaces, however, were created without due cognizance of the *anthropo-geographical* realities like ethnicity and culture, and their associated implications. Artificial national boundaries and alien economic structures were unilaterally imposed on the continent by the colonial powers, and these established a new set of internal and external relationships and their associated structures.

The International Relationships Established

As noted by Gore (1984), two types of international relationships linked to trade were established in relation to the colonial process, namely the relationship between the colonial powers in the form of inter-metropolitan trade on the one hand, and the relationship between the colonial powers and the colonies referred to as the "Metropolis-Satellite" and "Core-Periphery"

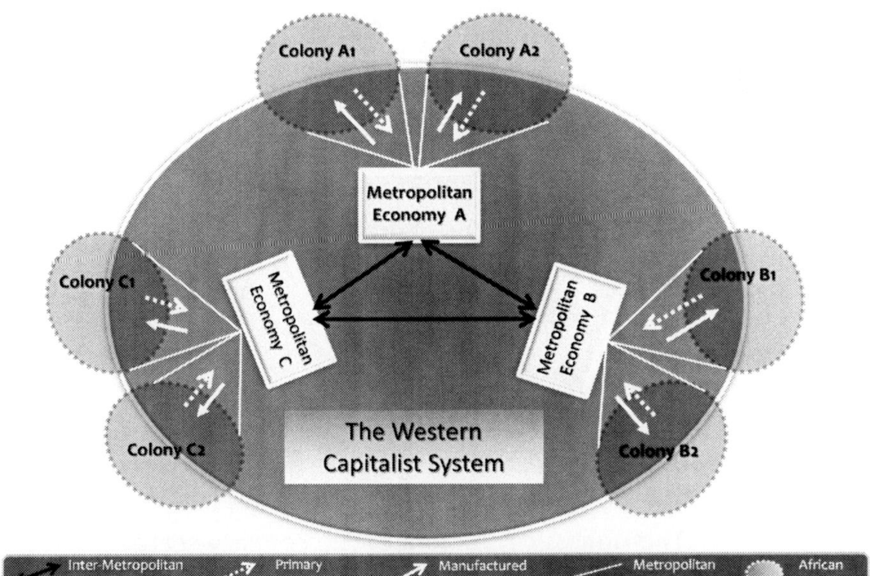

Figure 1.2. Metropolitan (Center)-Satellite (Periphery) Relations: Ambits and Trading Links.

relationships identified by Frank (1967) and Friedmann (1972) respectively, on the other. As figure 1.2 shows, both relationships operated within the parameters of the Western capitalist system but with different structural characteristics. The relationship between the metropolitan economies was relatively equal or symmetrical as reflected in the direct links that characterized inter-metropolitan trade among them. The relationships between the metropolitan economies and the colonies on the one hand (e.g., *A*, *B*, and *C* and A_1 and A_2; B_1 and B_2; and C_1 and C_2 respectively) and those between the colonies (e.g., between A_1 and A_2, or A_1 and C_1) were however very different. As noted earlier regarding the colonial experience, the creation of the colonies as capitalist "turfs" in Africa constituted the region's formal integration into the capitalist system but then this integration did not mean total absorption, as parts of the colonies continued to lie outside of the capitalist system, with relatively weaker state control, as shown by the segments outside the parameters of the Western capitalist system in figure 1.3. This is where our model departs from that of Slater (1975, figure 3.1) and that of Gore (1984,

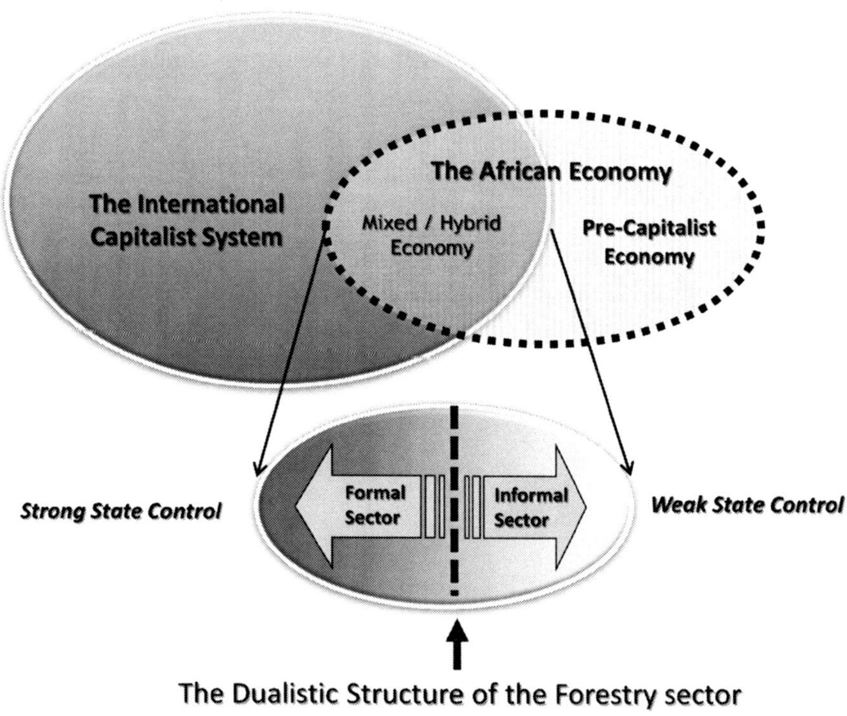

Figure 1.3. Formal and Informal Sectors of the African Economy
Source: Adapted from Owusu (2010).

figure 21). An important aspect of the relationships between the metropolitan economies and the colonies is that this external integration resulted in internal disintegration within the colonies. The integration established a rather unequal and asymmetrical relationship between the metropolitan countries and their respective turfs—the colonies—and each African colony tended to trade largely or exclusively with its respective metropolitan country as shown within the metropolitan ambits in figure 1.2 (see also, Slater, 1975).

As noted in Gore (1984, 131), "the key positions within the various colonial economies" tended to be "controlled by foreign metropolitan capital, . . . and commercial companies usually had monopolistic power. The colonial monetary system, with no exchange controls, allowed the 'siphoning of funds' from the colony, and the drain of 'surplus' was further promoted by the 'unequal exchange,' in Emmanuel's (1972) sense, of agricultural commodities produced in the colonies for industrial goods produced in the metropoles. Local capital accumulation was thus severely restricted."

A significant aspect of this Core-Periphery relationship was the colonial spatial structure that emerged. The new transportation systems that were developed were an integral part of the consolidation and extension of colonial control (Gore, 1984, p. 133). The road and railway networks generally extended from the coast to the rich resource areas inland to facilitate the export of raw materials to the colonial country and the import of manufactured items to the colony (Slater, 1975) and the mercantilist trade policies associated with colonialism also worked to suppress industrial development (Stutz et al., 2007, p. 61). In addition, as noted *inter alia* by Potter, et al. (1999, p. 33) regarding the internal processes associated with the colonialism in Africa, "The acceleration and *intensification of European capitalism brought about a breakdown of existing relationships between traders and African societies* into reaction" to the extent that Africa's colonial economies became so individually linked to their respective metropolitan economies as turfs, such that even today, genuine regional or subregional trade links continue to be inhibited (Sawyerr, 1990, p. 11). This is because each African colony, as a colonial turf, has tended to trade largely or exclusively with its respective metropolitan country (Slater, 1975) as shown in figure 1.2. It is with respect to these that Sawyerr (1990, p. 10) has argued that the objective effect of Africa's colonial experience was to reorder the social, political, and economic structures within each colony to integrate them into the world capitalist system. This was done, he notes, by aligning the dynamic sectors of each economy with the interests of the dominant classes of the metropolitan countries, and actually, creating within the colonies the political and social conditions necessary to sustain and reproduce the relationship established under the colonial system.

Contemporary Domestic and External Economic Structures

As shown above, the historical structure of relations between Europe and Africa is rooted in the diffusion of international capitalism during the colonial era as part of the turf creation process of capitalism. This structure was expressed in the form of unequal exchange in the area of trade, economic dependence, external control, and an unequal set of social arrangements within these turfs. In essence, therefore, the turf creation process meant that Africa, as noted by Sawyerr (1990, p. 28) had to "accommodate international capital" which "takes the form of rendering . . . labor and resources open to exploitation in situ; supplying raw materials and other inputs needed by the Western industrialized economies; and providing markets for their products."

For this purpose, turf creation, for international capital, ensures "unimpeded access to the raw materials and markets of Third World countries, as well as uninterrupted movement of capital into, and profits and rent *out of*, their economies." In fact, as Gwynne et al. (1999, p. 9) note regarding contemporary capitalist turfs, especially in this era of neo-liberalism and globalization,

> the impact of being more fully inserted into the global economy increasingly reduces the room for policy maneuver. In part this is because the governments of developing countries are more dependent on the policy approval of global institutions that "supervise" the world economy [such as the international monetary Fund (IMF), World Trade organization (WTO), and the World Bank] and on the investment decisions of multinational companies that can be swayed by the verdicts of international institutions.

The process of capitalist turf creation and the associated relational spaces created beginning with the colonial process are all new to Africa because pre-capitalist African societies were different from the capitalist European societies. Unlike the amoral economy—the impersonal marketplace which is concerned with economic transactions between members of the society—associated with capitalism, Africa's pre-capitalist societies were essentially moral economies. Exchanges defined personal relationships and social relations between members of the society. Thus, prior to the diffusion of international capitalism during the colonial era, Africa's pre-capitalist economies were characterized by an "economy of affection" in which economic action tended not to be motivated by individual profit alone, but was embedded in a range of social considerations that allow for redistribution of opportunities and benefits in a manner which is impossible in modern capitalism in which formalized state action dominates the process of redistribution (see Hyden, 1980).

As illustrated in figure 1.3, the process of incorporating the colonies into the international capitalist system led to the emergence of hybrid developing economies comprising a mixture of capitalist and pre-capitalist economies because African economies are to date, not fully incorporated into the international capitalist system as was shown in figure 1.2. African economies are therefore segmented into a continuum of two relational spaces, namely the formal and informal sectors. The formal sector, as a capitalist turf, is relatively more advanced with more state control, and involves all activities that are recorded as part of a country's official production, measured by the gross domestic product (GDP), the total monetary value of all goods and services officially recognized as produced in a country in a given year. The informal sector, on the other hand, encapsulates all economic activities—production of goods and services outside the formal sector. Thus, the informal sector operates largely outside the mainstream of government activity, regulation—less state control—and benefit, for work in this sector is often done for no cash payment or not reported to the government as taxable income.

Again, as shown in figure 1.3, the segmentation of the African economy into two relational spatial categories, formal and informal, indicates that the latter sector meshes into the former at the low end of a continuum, the highest end of which is advanced capitalism, and the lowest end being the pristine form of a pre-capitalist economy. Within national economies in Africa, intersegmental movements of factors of production and output are possible, in fact, frequent in the case of labor and goods, hence the broken line used in the diagram between the formal and the informal sectors (see Moser, 1978; Ninsin, 1991; Meagher, 1995; Yeboah, 1998). Indeed, it is within the space around the broken line that corrupt formal sector or public officials and bureaucrats collude with informal sector operators for mutual benefits or gains, especially since the former group tends to be largely underpaid, in a number of cases, or are just greedy, or take advantage of tempting situations for extra income, when accountability to the state or state oversight is minimal or nil.

The informal sector thus constitutes a turf within which capitalist and pre-capitalist elements struggle for control, or in certain cases, form alliances for mutual benefit as the case may be. In essence, the state is "suspended" to a degree over the informal sector (see Hyden, 1980), hence people, including some government officials, are able to move in and out of the informal sector to operate whenever deemed necessary for survival. This occurs particularly among the poor and unemployed; convenient; or personally profitable as in the case of some corrupt government officials (see Sarfo-Mensah, 2005). Unlike the informal sector in developed economies, where it is largely "underground" or illegal because it is outside the ambit of government control and regulation, it is openly acknowledged in SSA as an integral component of the overall economy, because it "plays a significant role in economic activity, particularly in production, distribution . . . and employment creation" (UN/ECA, 1986, p. 3).

As part of the struggle and control process, formal sector firms tend to exploit the informal sector at times, as a survival measure by capital, to control its own periodic crisis (see Weiss 1987 and Meagher 1995). This is because under the free market conditions, contractual relations will prevail to bind informal sector operators to formal sector firms, either for the supply of raw materials or semi-finished goods to the latter (Ninsin, 1991, p. 4; see also Moser, 1978). Although the informal sector renders useful and productive functions in African economies without formal controls and compensation, it is essentially a creation of the formal sector and in fact, in relational terms represents the failure of capitalism to transform African economies into successful industrial ones (Ninsin, 1991). It is in this regard that Potter, et al. (1999, p. 33) suggest regarding the internal processes associated with the colonialism in Africa, that "the acceleration and *intensification of European capitalism brought about a breakdown of existing relationships between traders and African societies* into reaction." A reflection of this reaction due to the breakdown of relations within the pre-capitalist economy, characterized by an "economy of affection," is the inequalities and asymmetrical relationships between the formal and informal sectors at the intra-national level. This is because the informal sector is, in fact, a dependent and subordinate part of the formal, and just as developed nations have devised mechanisms to exploit the developing ones, as noted by Ninsin (1991, p. 5), "in the case of formal-informal sector relations, the mechanisms ensure that formal sector operators or firms are able to accumulate at the expense of the informal sector operators because of the lower production and distribution cost that their links with the latter enables them to achieve."

The inherent inter-national and intra-national inequalities and asymmetrical relationships of the colonial system resulted in the struggle for independence after World War 1 (Davidson, 1994) and in 1957, Ghana, our case study and hitherto, Gold Coast, became the first SSA colony to achieve its independence under the leadership of Dr. Kwame Nkrumah. By 1965, almost all African countries had become independent as well. The remaining ones such as the Portuguese colonies of Angola and Mozambique and the British colony of Zimbabwe were among the last to become independent because of the need to fight the settler colonialists and their allies through devastating guerilla activities and other armed struggles.

Africa's structural problems became more apparent after independence. In fact, so successful was the colonial process of aligning the dynamic sectors of each economy with the interests of the dominant classes of the metropolitan countries, and creating within the colonies the political and social conditions necessary to sustain and reproduce the colonial relationship established, that the termination of formal colonialism—independence—nowhere effected a radical change in the relationship between the independent countries and metropolitan interests, at least at the level of the economy (Sawyerr, 1990, p. 10). Ninsin (1989), in relation to the structural problems, notes that as a result of the incorporation process, post-colonial governments have had two options with respect to relations with the Core:

a. to sustain the capitalist economy of the periphery together with its superstructural political apparatus dominated by the national bourgeoisie. This option demands a mandatory accommodation of international capital; and
b. to de-link or rupture the economic and other links with international capital. The purpose of this is to create a new structural basis and impetus for autonomous national development, i.e., negating the dictatorship of capitalist law of value through a rupture of the structural links that ensure such determination. This approach involves a radical restructuring of the old socio-political relations. However, as noted by Tandon (1985), with this option, there is a real prospect of collapse if it lacks a firm socialist vis-à-vis a mere popular content, given the resilience of the capitalist law of value.

In fact, it was in relation to the problems of colonialism and its enduring nature that Nkrumah emphasized Africa's need to de-link by avoiding "neo-colonialism," because where "the state which is subject to it is, in theory, independent, and has all the outward trappings of international sovereignty. In reality, its economic system and thus its political policy is directed from outside" (Nkrumah, 1965, p. ix). De-linking, it must be noted, does not mean autarchy. Rather, it is "the refusal to submit to the demands of the worldwide law of value, or the supposed 'rationality' of the system of world prices that embody the demands of reproduction of worldwide capital. It presupposes the society's capacity to define for itself an alternative range of criteria of rationality of internal economic options, in short a 'law of value of national application'" (Amin, 1990, p. 71).

Nkrumah's position in this respect is significant because the spatial structure of post-colonial Africa is largely "an unintegrated collection of non-industrialized, . . . non-self-sustaining and dependent states with little clout in international affairs" (Sawyerr, 1990, p. 11). This is because, as Sawyerr aptly sums up the colonial structural legacy,

1. political power is vested in an alliance of social classes with no fundamental interest in the transformation of the local political economy and a break from domination by metropolitan interests;

2. given its untransformed structures, and because of its one sided external orientation, the economy tends to work itself more closely into the international capitalist order, but in a position of subordination;
3. the majority of people, the principal victims of the continuation of colonial economic relations, are effectively disenfranchised and thus disabled to effect the necessary transformations; and
4. the economies are individually linked to the metropolitan economies in such a way as to inhibit genuine regional or sub-regional integration—a situation which persists, in spite of the good intentions, the rhetoric, the many meetings and the formal institutions for cooperation set up throughout SSA.

To address the colonial structural legacies, a variety of development models have been used. These range from the modernization approach, which appealed to African governments in the 1950s and 1960s, to the International Monetary Fund/World Bank structural adjustment programs, which became popular in the 1980s and 1990s but had become problematic by the beginning of the twenty-first century because it had failed, *inter alia,* to fashion spatial integration as a form of structural transformation.

SPATIAL INTEGRATION AS A GOAL OF STRUCTURAL TRANSFORMATION IN AFRICA

We noted earlier that structural problems in SSA include the export-bias of the region's economies, and the associated poor inter- and intra-sectoral linkages, which are all largely legacies of the colonial era (see UN/ECA 1986). We also defined structural transformation as the positive changes or progress made in the region's economies, resulting from national and international development policies aimed at correcting those imbalances. For our purposes, such positive changes must involve the emergence of effective or dynamic sectoral linkages (i.e., a strong functional interdependence) among manufacturing industries in the formal sector, and between other related sectors intra-nationally within the economies of SSA. Structural transformation in the region is indispensable because as pointed out by Fik (1999, pp. 20–21),

> a region's competitive position and relative growth potential is directly proportional to the density and strength of linkages found within its production network. In general, the stronger and more numerous the linkages within a region, both intra- and intersectorally, the more likely that region is to experience and sustain growth. Flows and linkages . . . are largely responsible for determining a region's economic status.

What then is the nature of a spatially integrated economy necessitating it as a goal of structural transformation in SSA? As aptly defined by Mabogunje (1981, p. 62) a spatially integrated economy "has a certain wholeness to it which enables policies applied to any part of it to have system-wide effects or repercussions. Such an economy is intrinsically centered on itself and it is not a composite of parts, some of which are autochthonous while others are more effectively involved in another and exterior spatial system." This definition is rooted in the "relational" conceptualization of space, discussed earlier, which interrelates rather than separates spatial patterns and social processes. In this sense, SSA economies, as capitalist relational spaces, have been shaped, and in fact, acquired their respective specific forms and styles of disposition largely as a result of capitalism (see Mabogunje, 1981, p. 52). As a goal of structural transformation, therefore, the core of SSA's spatial integrative relations should comprise the existence of an effective set of linkages between different places and activities in the region. *Inter alia,* these relations must be a function of both the geographic flows of material, and the vertical relations in the production process especially between primary, processing, and distributive activities (see Mabogunje, 1981, p. 61). The associated distributive activities, thus, extend beyond SSA in the form of external or international trade.

Structural transformation aimed at effecting spatial integration is tied to the view that at the core of underdevelopment in SSA is the lack of fundamental sectoral linkages necessary for spatially integrated development in the region (Darkoh, 1971; Mabogunje, 1981; OAU, 1982; Sawyerr, 1990). This is a result of Africa's colonial experience, as noted earlier, that re-ordered the economic structures within the region's colonies to integrate them into the world capitalist system, whereby the dynamic sectors of each economy were aligned with the interests of the metropolitan countries (Sawyerr, 1990, p. 10). In acknowledging this problem, the United Nations Economic Commission for Africa (UN/ECA) noted that "the structure of the African economy defines the essential features of Africa's central problem of underdevelopment" (UN/ECA 1986, 1). This was based on its observation that the general post-colonial structure of the region's economies had continued to be "colonial."

Structural transformation via spatial integration is therefore basic to three development processes in Africa: 1. Changing the region's economies from their essentially colonial characteristics, 2. Forging the crucial "wholeness" that characterizes advanced economies, and 3. Making them intrinsically self-centered although by no means autarchic (see Amin, 1990). It was in recognition of this that the development of spatially integrated economies, which is incompatible with the accommodation of international capital, in the region was expressed as a desired goal by both the Organization of African

Unity (OAU) and the United Nations Economic Commission for Africa (see OAU, 1982; UN/ECA, 1986). Indeed, it was to engender such transformation that African heads of state and of government prescribed the Lagos Plan of Action (LPA) in an OAU summit meeting in 1980. The signatories of the LPA emphasized "the creation of an industrial structure at the national levels within the framework of an integrated economy" (OAU, 1982, p. 21), as part of the Plan's long-term objectives (i.e., by the year 2000) at that time. This emphasis derived from the importance of manufacturing industry in the development process. For the peripheral SSA countries, the importance of manufacturing industry with its associated spatial structure lies in its dynamic function as the "heart of the economy given its radiant and repercussionary effects on the other sectors of the economy" (Kayode, 1987, p. 145).

It was also in relation to the colonial structural vestiges and the World Bank-IMF prescribed Structural Adjustment Programs (SAPs) to redress the structural problems in the region, that the OAU held a summit meeting in 1985 devoted primarily to the discussion of the African economic crisis. The meeting produced an important document, *Africa's Priority Program for Economic Recovery 1986–1990* (OAU, 1985), in which the African leaders lamented that five years after the adoption of the LPA, little progress had been made towards spatially integrated development. Among the obstacles that were identified, were (a) the failure to restructure the inherited colonial economies; (b) the deepening of dependency through the misallocation of domestic resources and over-reliance on external financial resources, and (c) the very slow development of co-operation and integration in new sub-regional groupings (see OAU, 1985; Sawyerr 1990, p. 17). The leaders' frustrations are all tied to human performance—the Human Factor (HF) in the development process.

THE RELEVANCE OF THE HUMAN FACTOR IN AFRICA'S DEVELOPMENT

Ultimately, the contribution of the forestry sector and its associated timber industry to structural transformation in the development process in Africa, in general, is tied to human performance, which is a key element in the structuralist perspective of this book. It is in this respect that the role of the HF in development concept comes to bear. Adjibolosoo (1993, p. 142) succinctly defines HF as "the spectrum of personality characteristics and other dimensions of human performance that enable social, economic and political institutions to function and remain functional, over time." HF in development comprises two elements. The positive aspects of HF sustain the workings and application of the rule of law, political harmony, a disciplined labor force, just legal systems, social welfare, etc. The negative personality characteristics and other dimensions of human performance such as corruption, dishonesty, lack of commitment to selfless and patriotic efforts, and to accountability, constitute what is regarded as HF decay. This includes issues of corruption, and the misuse of state power, such as the adoption of "*Faustian deals,*" namely agreements—at both the international or domestic levels—which are contracted for present gain without regard for future cost or consequences. Given that, no social, economic, or political institutions can function effectively without being upheld by a network of committed persons who continually stand firmly by them, as they work to engender development (see Adjibolosoo, 1993, p. 142), it is important that the critical role of HF to the development process is tied to the post-LPA development obstacles that were identified by the African leaders. These obstacles and their emergence are reflected in the spatial and organizational structure of Ghana's timber industry. To put these in perspective, chapter 2 focuses on the historical development of the industry as an expression of the dynamics of capital in the creation of turfs based on resource exploitation, the role the industry plays in the Ghanaian economy in terms of how such turfs are organized internally and externally, and the nation's development process in terms of spatial integration with respect to structural transformation in SSA.

Chapter Two

Turf Establishment and the Development Process

Structural Foundations of Ghana's Timber Industry

This chapter presents the historical foundations for our discussion. It focuses on the development of Ghana's timber industry, in terms of the evolution of its structure and organization, and the role it plays in the Ghanaian economy as well as the nation's development process. In doing so, it illustrates and discusses the creation of a capitalist space or turf in Africa, based on the exploitation of timber, and shows the beginnings of the nature of the tropical Africa's timber trade internally and externally. In this respect, pertinent colonial policies regarding the timber industry are discussed. Situating the process of capitalist turf creation and the associated structural patterns in a geographical and historical perspective is very important since geographic space—absolute, relative, and relational—and development processes cannot and should not be divorced from each other, for neither can exist independently of the other (Sayer, R. A., 1985, p. 5; Gore, 1984, p. 181). Consequently, as pointed out by Gore (1984, p. 144), "spatial patterns in developing countries must be analyzed historically and (should) take account of the dependent position of those countries in the international economic system," as suggested by Slater (1974, 1975), Dos Santos (1969), and Santos (1979).

TROPICAL TIMBER, TURF CREATION, AND SPATIAL INTEGRATION

What are some of the socio-economic importance of tropical timber that make an industry based on it worthy of analysis in terms of capitalist turfs, and the development process in Sub-Saharan Africa in general and Ghana in particular? In general, the tropical timber industry constitutes a major source of income and integration of products and services for both the tropical timber producing sub-Saharan African economies as well as for the global economy as a whole. According to Switzerland's State Secretariat for Economy (SECO), 190 million m^3 of tropical timber are produced each year worldwide, of which 40 million are exported (SECO, 2004). Regarding capitalist turfs, controls on the sources and organization of production as well as the size and dynamics of the markets are critical for capital accumulation. This is because tropical timber constitutes a major component of international trade in the global economy by virtue of the relatively unique industrial and aesthetic qualities of the various types in terms of their consumption. Table 2.1 shows the main end uses of tropical timber in some of the major consuming countries in the global economy.

The significance of these end-uses is that each one has its own and related production networks and activities—spatial integration, with significant social and economic implications for the consuming countries. The greater the network, the more intensive the linkages, and the robustness of the associated production activities and consumption, the greater the socio-economic and even geopolitical benefits reaped by the consuming country. In that respect, even Communist China, bitten by the capitalist bug, is carving out a turf for herself in Africa. Carmody and Owusu (2007, p. 506) for example, pointed out that the Chinese state has identified "Africa as a strategic economic space"—a turf—and that part of China's geo-economic strategy is to ensure access to critical natural resources to maintain its economic growth. Also, Butler (2005) in her piece, "Growing Pains and Growing Alliances: China, Timber and Africa—Timber hungry China moves into Africa," observes that *inter alia*, "China, as the fastest growing economy in the world, is poised to make significant impacts on the global market . . . , especially with its expanding involvement with nations rich in natural resources but deficient in economic and political stability. Nowhere is this more apparent than in Africa where China has rapidly bolstered its ties in recent years with the majority of the continent's fifty-four nations. . . . Among other resources, the continent is becoming a major global supplier of timber and China may soon control access to this resource." She further notes that "Beyond establishing control

Table 2.1. End-Uses of Tropical Timber in the Major Consuming Countries

Form	*End-Uses*
Lumber	Furniture, flooring, windows and doors (both external and internal), staircases and other items of joinery, garden furniture and decking, marine and sea applications, and transport.
Veneer	Plywood manufacture, laminating medium density fiberboard and particle board, and furniture.
Plywood	Shop fittings, kitchens, furniture, building, and transport.

Source: FAO, 1999

over African resources, China is looking to line up economically marginal nations to serve as producers of and markets for its products in the future." This is because China has developed an insatiable demand for resources as a result of its tremendous economic growth, and that "with growth rates exceeding 7%, it is estimated that by 2020 the GDP of China will exceed that of Japan, behind only the US and EU." China's need for resources compels it to pursue resources beyond its borders, she argues.

Canada and the United States have traditionally been relatively small participants in the tropical timber trade and in 1990, their combined import value of tropical logs, lumber veneer, and plywood totaled about US$550 million. North American imports of tropical timber are supplied by Indonesia, Malaysia, and Brazil, with imports from Africa relatively negligible (Barbier et al., 1994, p. 39). Unlike the Chinese, the United States has taken a combination of different approaches within its neo-liberal policy framework to stake turfs in Africa. During his trip to Africa in 2003, President Bush referred to Africa as the "last great emerging market of the world" (see Carter III, 2004) and the African Growth and Opportunity Act (AGOA), which "offers tangible incentives for African countries to continue their efforts to open their economies and build free markets" (see AGOA, 2000) constitutes one of the mechanisms used to help form linkages that will tap into this market and create new investment opportunities over the long term in the region.

For the timber-producing countries or economies of SSA, whilst meeting some of the basic economic and other needs of the people, such as employment, construction, and fuel needs, it also contributes to their respective gross domestic product (GDP) and is potentially, a means for poverty reduction and structural transformation to promote spatial integration. For example, in the case of Ghana, the industry is one of the most important sectors in the domestic economy. With a natural forest resource base of 109,000 km^2, vis à vis West Africa's total of 615, 000 km^2 as at 1980 (WRI, 1994), the sector contributes some 6% of the country's total GDP and 11% of total commodity export earnings annually. The forest with its wildlife resources also provides direct employment to over 100,000 people and indirect employment to over 2.5 million people (Forestry Commission, n.d.), while 60–70% of the population also depends upon the forest resources for their livelihoods and cultural purposes (Arthur et al., 2007). In addition, 14 million m^3 of fuel wood from the forest, account for 75% of the country's energy needs (Forestry Commission, n.d.). In fact, over the years, wood has been the third most important component, after cocoa and minerals, among commodity exports (see TEDB, n.d., p. 1). In 1994, the wood industry contributed a record high of US$230 million in exports to the nation's foreign exchange earnings. Subsequent annual export earnings from the industry averaged US$180 million (*FAO, II*).

Current export earnings from forests and wildlife are reported to be approximately US$400 million per annum, and whilst environmental benefits from the forest cannot be quantified, they include the provision of ecological stability, watershed protection, and a reliable source of food, medicinal plants, and building materials (University of Ghana, n.d.).

Ghana's timber industry can thus be used to test the evidence of structural transformation in Africa from the colonial period through the neo-liberal era at the beginning of the 21st century, in terms of the emergent intra-national (i.e., intra- and inter-sectoral) and international linkage patterns, for several reasons. First, it is a key sector that links the primary and secondary activities within the Ghanaian economy. Second, its post-colonial linkage structure has long been identified as a development problem (Combé, 1972; Owusu, 1995; Forestry Commission 1988). Third, it is a manufacturing industry with immense potential for dynamic upstream and downstream linkages based on a renewable resource. Also, it has the potential for lateral linkages. Hence it is capable of playing the crucial and dynamic integrative role in the development process as underscored by Kayode (1987). This last attribute is very important because the changes in the industry must suggest progress towards a nationally integrated economy as envisaged under the LPA,[1] for example, which specifically stressed "*the need to establish links between industry and other sectors as well as between various industrial sub-sectors, so as to promote interdependence among them and achieve harmonized industrialization and overall economic development*" (OAU, 1982, p. 21).

To show the post-colonial linkage structure that has long been identified as a development problem, let us put the development of the industry in a geographical and historical context to establish the absolute, relative, and relational spaces regarding the industry, for our discussion.

EVOLUTION OF THE PROBLEM: THE GEOGRAPHICAL AND HISTORICAL CONTEXT

The Republic of Ghana (formerly, Gold Coast until its independence from the British in 1957) is a relatively small country located on the West Coast of Africa. With a total land area of about 91.8 thousand square miles, it is located about 489 miles north of the Equator on the Gulf of Guinea, between latitudes 4° and 11° North (see figure 2.1). It is bounded on the north by Burkina Faso, on the west by Côte d'Ivoire, on the east by Togo, and on the south by the Atlantic Ocean.

As a tropical timber-producing country, Ghana has a climate characterized by moderate temperatures generally between 70° F and 90° F and a rainfall

Figure 2.1. Location and Regions of Ghana

Used by permission of the Cartographic Section of the United Nations organization.

regime that facilitates the growth of the timber. There are the two rainy seasons occurring between March and July and from September through October, separated by a short dry season in August and a relatively long dry season from mid-October through March in the south. Of the total land area, 31.8 thousand square miles in the southern sector falls within the rich forest belt—the Tropical High Forest or Closed Forest Zone which constitutes the source of Ghana's commercial timbers (see figure 2.2).

The Tropical High Forest or the Closed Forest zone comprises the Rain Forest, which covers the southwestern section of the country and less than 10% of the zone, and the Semi-Deciduous Forest, which makes up the remaining 90% of the zone. It is the Semi-Deciduous Forest which contains the majority of the commercial species, the popular ones among which include Wawa (*Triplochiton scleroxylon*), Mansonia (*Mansonia altissima*), Danta (*Nesogordonia papaverifera*), Mahogany (*Khaya ivorensis*), and Makore (*Tieghemella heckelii*) (ITTO, 2005, 98). The High Forest Zone, which used to cover 34% of the land area by the late 1980s, now covers about 29.35% of total land area as a result of deforestation. Between 1990 and 2000, the average annual estimated rate of deforestation was 120,000 hectares, with the average growing stock in the remaining forest area being only 49 m^3 per hectare (ITTO, 2005). The annual rainfall in the zone averages 81.2 inches, but varies greatly with the heaviest rainfall in the Western Region. This rainfall regime also affects the road conditions, and therefore the costs, for the transportation of wood from the forest for the industry. This is particularly so during the rainy season, when most of the roads become virtually unusable for vehicles. Stretches along the coast and the northern sector of the country are drier and the predominant vegetation is savannah and scrub, which now covers 70.65% of the country, up from 66% of total land area previously.

It is in relation with this physical geographical (i.e., absolute and relative spatial) context that the development of the industry and its associated relational spaces evolved over time[2] and it derived primarily from the country's incorporation into the global capitalist system (Frank, 1969; Wallerstein, 1974; Slater, 1975; Andrae, 1981; Sawyerr, 1990). As a peripheral economy within the global market system, Ghana's transition from a pre-capitalist economy into a monetized one began in the fifteenth century. The year 1471 marked the first contact of Europeans (Portuguese mariners) with the people of present day Ghana.[3] The contact initiated a trade involving the exchange of gold (hence its former name Gold Coast) for such items as firearms, ammunitions, rum, and cloth. In time, these were superseded by the export of slaves and subsequently, by such primary commodities as palm oil, cocoa, and timber, particularly after the British colonized the coast in 1874 and the colonization of the Ashanti Kingdom which extended across the high forest zone in 1902.

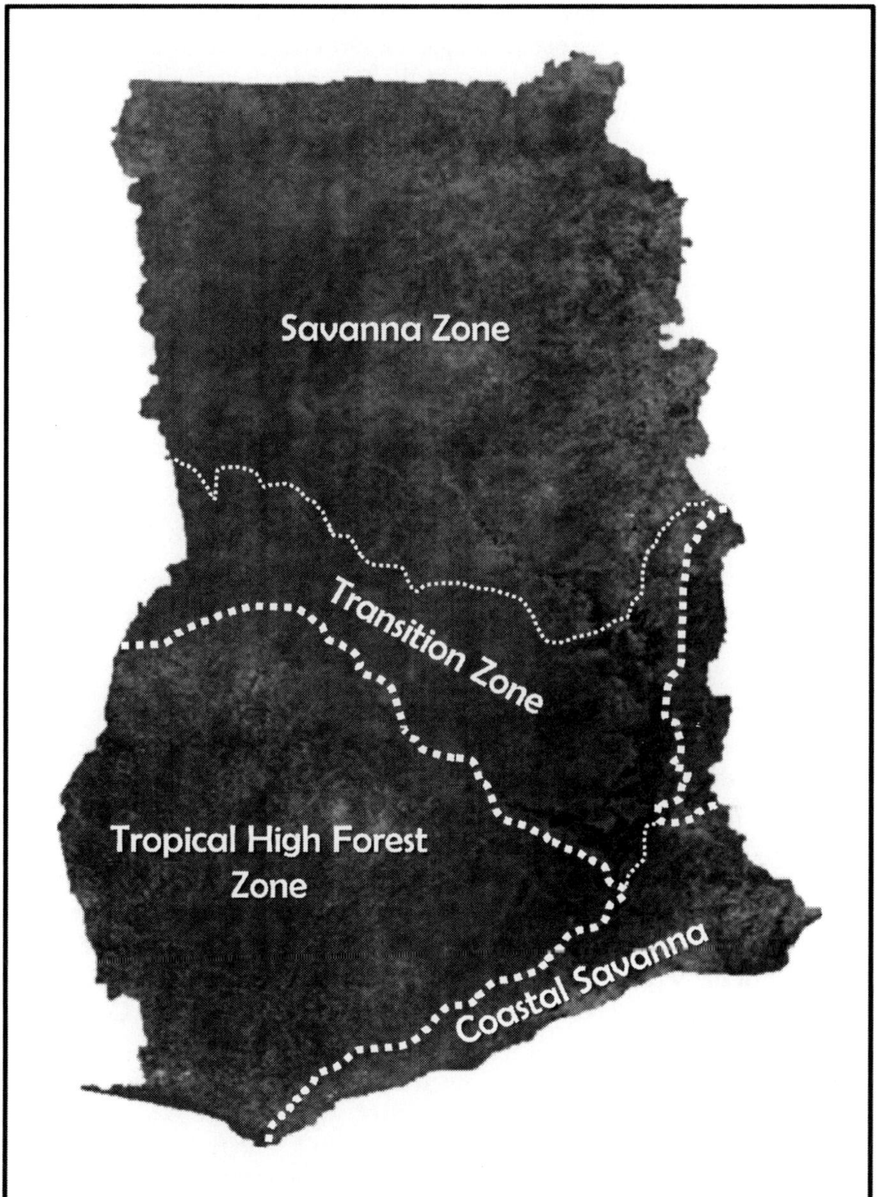

Figure 2.2. Ghana: Vegetation Zones
Source: Adapted from Wikipedia (2006)

In relation to this, the FERN Report (2006) argues that, with respect to the relational space and the associated capitalist turf created, Gold Coast as a colonial state was essentially a creation of the European merchant houses. As a colony, it was created by international capital to supervise the final destruction of pre-colonial modes of production based on communal agriculture, low-volume manufacture, and trade (including trans-continental trade in salt, gold, leathers, etc.) that survived the trans-Atlantic slave trade. In terms of the socially created turfs, the Report notes that the colonial process subordinated the pre-colonial elites through military conquest or diplomacy, and reduced them to paid "Native Authorities" within a colonial local government system. The British then enhanced or created Native Authorities' coercive powers over their communities in order to facilitate foreign corporate access to landed resources, and to develop through conscription, the infrastructure necessary for an export economy—roads, railways, mines, ports, and so on. This was then followed by

1. an introduction, or encouragement of the development of, cash crops for export; and
2. The discouragement of traditional subsistence economic activities in a bid to force communities into the cash economy and to cut off their links with their resources.

These facilitated the establishment of a dependent "national" economy based upon export of minerals, cash crops, and timber to Europe, in exchange for cash with which to import industrially manufactured commodities. More significantly, the colonial state was then able to guarantee investors cheap access to mineral and timber concessions through the Native Authorities. These measures also enabled them to ensure low taxation of extractive industries while the workings of the "international market" ensured that peasant-produced commodities were available at less than their value. It also ensured that most of the wealth produced in the colony was realized and retained by European companies. The residual wealth remaining in the Gold Coast went to pay the cost of the colonial administration and to reward local collaborators; almost nothing was left over for the rural communities who depended on and legally owned these resources (FERN, 2006, p. 19).

In relation to international capital's strategy to facilitate foreign corporate access to the colony's resources by developing the infrastructure necessary for an export economy, Szereszewki (1965, p. 1) for example, points out that the process of economic development that took place in the Gold Coast between the years 1891 and 1911 fundamentally changed the structure of the economy. He notes:

> A railway network was in existence. Whole new sectors of activity appeared during the two decades, and by 1911 the structure of the economy was trans-

> formed. What is even more interesting, the transformations of this period largely determined the structure of the economy for the next fifty years, and it can be maintained that the pattern established in 1911 still largely persisted in 1960 (Szereszewki, 1965, p. 2).

By the end of the period then, the outline of the economic geography of present day Ghana, based on the railway-connected triangle Accra-Kumasi-Takoradi (then Sekondi) was clearly visible. With regards to the timber industry, Szereszewki noted that the evidence of a sawmill in the Sekondi District, and the expansion of the European logging firms which obtained 205 m^2 of concessions in the colony between 1906 and 1911 were an indication of the industry's importance in the structure of the economy then as it is today. This pattern continues largely to endure even at the beginning of the twenty-first century.

DEVELOPMENTS IN THE TIMBER INDUSTRY: FROM THE PRE-COLONIAL TO THE NEO-LIBERAL ERA (1800–2000)

Six distinct phases of change can be identified in the development of the industry. Phase One covers the genesis of the industry in 1887 through the end of the First World War (WWI). Phase Two covers the inter-war period; and Phase Three, the beginning of the Second World War (WWII) through Ghana's attainment of independence from Britain in 1957. The Fourth phase spanned the period covering 1957 through 1966, when the administration of the first post-colonial government of Ghana was ousted; and the Fifth phase marked the period thereafter through April 1983, when neo-liberal policies in the form of the World Bank-IMF free market based structural adjustment dubbed the Economic Recovery Program (ERP) was adopted. The ERP thus constitutes the Sixth phase. These phases, for our discussion purposes, are broken into two eras, subsuming the first three and the remaining three under the Pre-Independence (Pre-Colonial and Colonial) Era, and the Post-Independence (Post-Colonial and Neo-liberal) Era respectively. Each era was driven by certain external and internal factors and policies. For convenience, we will consider the Pre-Independence Era in this chapter, and the Post-Independence Era in Chapter 3.

PHASE I (THE PRE-INDEPENDENCE ERA): 1887–END OF WORLD WAR I

The Core-Periphery relationship and spatial structure with regard to the timber industry began to be established in this phase. The formal genesis of Ghana's timber industry and its linkage to Europe was in 1887, when 240 superficial cubic feet of Mahogany (*Khaya*) and Utile (*Etandrophragma*) were first exported to Britain as samples on an experimental basis, thus introducing them formally on the world market and initiating Britain, and for that matter, Western Europe as the principal consumer of Ghana wood. According to Dickson (1969), this was however preceded by a short period of timber exports a little before 1800, at which time timber was exported from a small port at Axim on the coast of the present day Western Region (see figure 2.1) and a few neighboring small ports. By 1800 however, the immediate vicinity of the ports had been denuded of exportable timber. Exports then practically ceased as a result of technological difficulties related to extraction and transport since it had become necessary to transport timber over long distances from the forest to the ports but the absence of wagons, tools, and machinery for working the wood made the task too formidable. Timber logging resumed in 1887 because Britain now controlled much of southern Ghana, and therefore there was no longer any political barrier to use the Tano, Ankobra, and Pra Rivers as the principal modes for transporting the timber from the forest to the coast.

As an integral part of Britain's consolidation and extension of her colonial control as pointed out by Gore (1984), Britain began to construct its first railroad from Sekondi to Tarkwa in 1898 and completed it in 1901, which immediately affected the volume of timber exports, with environmental consequences (see figure 2.3). In 1906, for example, there was a drop in exports

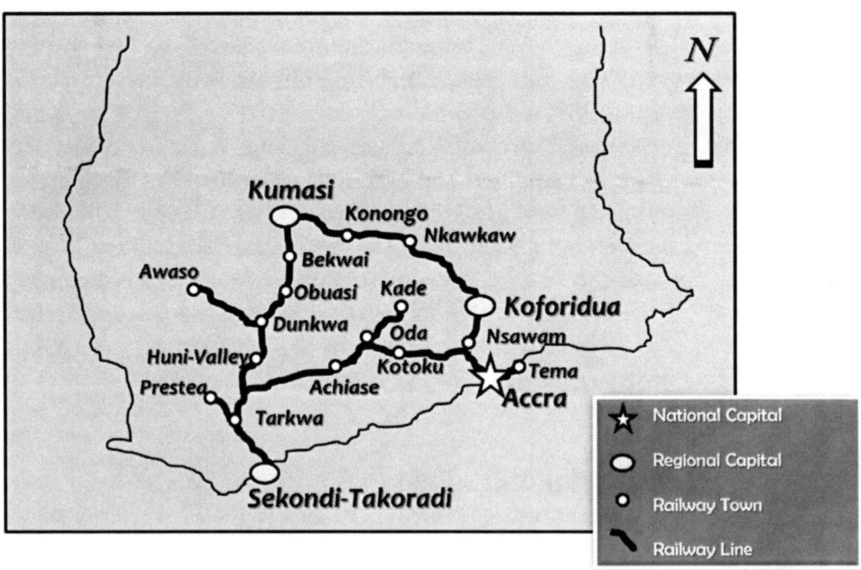

Figure 2.3. Ghana Railway Network

because there was a near total elimination of marketable timber near the Sekondi-Tarkwa railroad because of the accessibility the line offered. At the rate which the timber was being extracted, it became necessary in 1907 for the colonial government to enact a Timber Protection Ordinance prohibiting the felling of immature trees of certain species below certain specified girths for export. In 1909, a Forestry Department was established which gave high priority to the creation of forest reserves. This was because the Department held the view that forest reservation was an essential requisite of its program, and by its 1927 Forest Ordinance Act, the then Guggisberg government regularized the timber industry and further protected forests from overexploitation (Adu Boahen, 2000, p. 112). This was linked to developing the cocoa industry as the department sought to engage in forest reservation with the objective of enabling it to "maintain the high atmospheric humidity necessary for the cocoa industry, to which the dry North-East wind or *harmattan* presents a constant threat." Significantly, the department's Forest Policy of 1948 then sought to preserve the physical environment, even further, to ensure maximum productivity and value on the basis of sustained yield, and the conduct of scientific research. It was also to control forest resources to ensure prolonged supplies and conservation (Boateng, 1991).

Land Ownership Structure and the Concession System

The industry's genesis coincided with the establishment of another raw material extraction process—cocoa farming, which involved vegetation clearance, in the closed forest zone. Exports then increased so dramatically that by 1894, as much as 450,000 cubic feet of timber were exported, increasing further, to 3,000,000 cubic feet by 1913, thus firmly establishing the industry. Numerous timber concessions were then taken by foreign capitalists, thus initiating the subsequent pattern of using concessions in the industry. Labor for both activities was attracted from both inside and outside the country (Dickson, 1969, pp. 176–77, Adu-Boahen, 2000, p. 93).

Land ownership structure in the Gold Coast was a very complicated one. Essentially, however, lands and the forests thereon were communally owned and were thus vested in various tribal "stools" which traditionally constitute the framework in which the individual or family could claim access to natural resources. Timber exploitation from the forests by companies and contractors had to be done through agreements with the stools in the form of concessions. Two main types of concessions were identifiable: 1. The License, which conveyed only the right to harvest the timber, with no proprietary rights in the land itself; and 2. The Lease, which however, was the more frequent type of arrangement. It involved proprietary interest in the land which is granted for a period by the owner to the tenant, usually in consideration of an annual rent (Gordon, 1966). Because of the complicated ownership of stool lands, the British never stipulated the allocation of concessions in a forestry law. Instead they relied on a special Concessions Ordinance of 1900 which was mainly for forest management, as an easy solution for organizing the exploitation of forest areas which were otherwise inaccessible. However, as Combé (1972, pp. 14–15) points out, such a system has the disadvantage of avoiding meaningful contact between the owner of the forest and the timber producer. It was an easy solution for the government because it made it possible to manage the huge forest areas with a small administrative staff, and the allocated forest areas did not require any capital investment by the government. Communications, equipment, and specialists must be paid for by the concessionaire.

Both the lessees and licensees had to pay royalty per tree at a rate based on the specie. An annual rent also had to be paid based on the square mileage of the concession and a sylvicultural fee based on acreage had to be paid by the leasees. Both types of concessionaires had to pay compensation for each fruit-bearing tree or cocoa tree damaged by the contractor. Furthermore, they had to pay a fee to register their property marks (i.e., the company identity symbols) which had to be renewed annually. This concession system, however, had three main disadvantages. First, it favored the uncontrolled exploitation and destruction of the forest, since the concessionaires tended to be mainly interested in the maximum financial rents from their operations. Secondly, the allocation of large areas gave few entrepreneurs monopolistic control, resulting often in limited competition on the market.

Finally, contractors could keep their concessions without extracting the timber potential for speculative reasons and consequently retard the economic development of the area (Combé, 1972, p. 17). It was to address the main disadvantage namely, the uncontrolled exploitation of the forest in the wake of the increasing number of foreign capitalists, that the colonial government wisely forbade the extraction of timber below a certain girth through the Timber Protection Ordinance of 1907. The control of the industry by foreign capitalists also initiated the pattern of foreign domination in the ownership of firms in the extraction and primary-stage processing in the industry. This was because it was these foreign capitalists *vis à vis* the indigenes, who had the capital to engage in such operations at that time.

In terms of trade, the volume of timber exports in the first phase of the industry's development, particularly after the 1890s, depended partly on demand factors in Western Europe and the United States, and partly on lack of transport facilities in Ashanti and the eastern and central regions of the colony. In the case of transport facilities for example, the regimes of the Tano, Ankobra, and Pra Rivers partly determined the volume of timber available for export. Periods of flood facilitated the export of larger quantities as in 1907 and 1912, and a vastly reduced volume when they were particularly dry as in 1895. As noted earlier, exports reached the unprecedented volume of 3 million cubic feet in 1913, but dropped to less than half of that volume during WWI. As pointed out by Dickson (1969, p. 177) demand declined with the virtual shutdown of the cabinet-making trade in Europe.

As noted earlier, the first railway line from Sekondi to Tarkwa was begun in 1898, and reached Tarkwa in 1901 (see figure 2.3). It was extended to Obuasi, the rich source of gold in 1902, and the political and commercial center of the Ashanti kingdom in 1903. A branch of this line, the Tarkwa-Prestea line, was built between 1908 and 1911, and the Accra-Kumasi line, built mainly for the development of the cocoa industry, was begun from Accra in 1900, though not completed until 1923. It must be noted that the construction of the railways was a solution to the ineffectiveness of the existing infrastructure such as the dirt roads both for conveying heavy machinery needed for mining of such resources as gold and diamonds in the interior and for transporting heavy metal ores like manganese and bauxite to the ports. Adu Boahen (2000, p.101) aptly sums it up this way:

> By 1911, the structure of the economy had undergone complete and fundamental change. The most noticeable characteristic . . . at this time was the neglect of industrialization, overdependence on the cocoa crop, and the total domination of mining, banking, and the foreign trade sectors by expatriate firms. In short, the economy that had emerged by the beginning of (the) century was a colonial economy marked by the exploitation of the natural resources of the country on the part of the colonial power and her trading firms to the advantage of European and British shareholders abroad and not to that of the Ghanaians themselves.

The process of consolidation and extension of colonial control in the Core–Periphery relationship between Britain and Gold Coast, and the associated colonial spatial structure that emerged continued in the second phase—the inter-war period. The road and railway networks continued to be extended from the coast to the rich resource areas inland to facilitate the export of raw materials to Europe and the import of manufactured items to the colony (see Rodney, 1972; Slater, 1975).

PHASE II (THE PRE-COLONIAL AND COLONIAL ERAS): THE INTER-WAR PERIOD (1919–1939)

Regarding the lack of transportation facilities in the Ashanti, the Eastern and Central regions of the colony, which partly affected the volume of timber exports in the first phase of the industry's development, Adu Boahen (2000,

p. 93) notes that it was not until the completion of the Accra-Kumasi railway in 1923 and the Huni Valley-Kade railway (see figure 2.1) in 1927 that the industry revived. This was because as noted by Dickson (1969, p. 178), the upper reaches of the rivers in Ashanti were not free of obstacles, while the rivers flowing through the forest in the Central and Eastern Regions were not as large as those in the Western Region. The Volta River was too far from the exploitable areas and so the only alternative was by land and that was extremely difficult. It was therefore with the opening of the Accra-Kumasi line in 1923 and the Huni Valley-Kade line that it became possible to develop the industry in the Eastern and Central Regions. The network terminals at Kumasi (Ashanti), Sekondi-Takoradi (Western), and Accra (the national capital) thus contributed to initiating the eventual concentration of firms in those regional capitals. The establishment of this railway network also initiated the pattern of the industry's initial dependence on the railway system, as well as a contributor to the spatial pattern whereby wood-processing firms like sawmills tended to be concentrated in urban/metropolitan centers like Kumasi and Takoradi.

Again, regarding trade, there was another decline in demand in both Europe and the United States in this phase, during the depression years of the early 1930s, and that severely reduced exports to an annual average of some 500,000 cubic feet. This demonstrates the vulnerability of the industry to external demand. Hitherto, the bulk of the trade was in the form of logs, but sawn timber became popular during this phase. Table 2.2 shows a preliminary census in 1939 regarding the approximate distribution of pit-sawn timber production.

In addition, 140,000 $ft.^3$ or so of mill sawn timber was produced partly for export by the fifteen establishments whose equipment varied from a simple circular saw to more elaborate machinery which existed at that time. The two largest sawmills owned by the British, were located in the Western Region and operated by Messrs. Thompson, Moir, and Galloway and the West African Mahogany Company. These set the precedent of establishing very large

Table 2.2. Regional Production of Pit-Sawn Timber (1936–1937 to 1939–1940)

Region	*Volume ($Ft.^3$)*
Eastern	50,000–100,000
Central	100,000
Western	750,000
Ashanti-Brong Ahafo	270,000

Source: Department of Forestry, Annual Reports, 1936–1937 to 1939–1940. Quoted from Dickson (1969, 319).

companies in the Western Region. The establishment of sawmills was in no way inconsistent with the colonial policy suppressing industrial development as noted by Stutz et al. (2007, p. 61) because lumber essentially constitutes a raw material that is used for further processing or finished products in Europe. It was late in this period that Gold Coast wood was successfully advertised at the Empire Exhibition in Glasgow in 1938 and thus increased the prospect of exporting other species of mahogany as well as new ones including Odum (*Chlorophora excelsa*) and Wawa (*Triplochiton scleroxylon*).

Evolution of the Domestic Market

The nature of the domestic market during this phase laid the foundation for the patterns whereby most of the processed and high quality wood were exported to Europe. This left the relatively inferior processed wood and rejects from the sawmills for consumption on the domestic market. Although production for domestic consumption was important at this time since timber was needed for a variety of purposes, and despite the fact that there were many varieties of hard wood available, the Odum (*Milicia excelsa)* was the most fancied for all purposes. That restricted, to a great extent, processing of that species for the domestic market because there was a deeply rooted conviction that it was the only timber of any value (Dickson, 1969, p. 319). For soft wood however, Boateng (1991) shows that prior to WWII, the country depended on imports for its major lumber requirements. This was however, curtailed and soon ceased with the beginning of the war. The domestic market was also not very reliable in terms of disposing of the processed wood produced. This was because production was subject to fluctuations related to the nature of the cocoa market. Producers were better off focusing on production for the export market, thus contributing to the initiating of the pattern whereby most of the processed wood was exported in the pre-independence era. The reason was that, if the price of cocoa was high, it resulted in a big diversion of labor to the easier and better paid work of harvesting cocoa. Consequently, timber became scarce relative to demand on the domestic market and its price nearly doubled. The reverse was the case when the supply increased as a result of the availability of increased supplies of labor driven to the timber camps by a drop in cocoa prices (Dickson, 1969, p. 319).

In part therefore, the establishment of the large sawmills in the Western Region and subsequently, elsewhere, was to stabilize production and that required that large capital must necessarily be invested so as to maintain permanent and competent gangs of workmen to keep the mills running. It was export-orientation that could sustain such operations. The pattern of reliance on capital-intensive methods of operation that existed in the major sawmills prior to the colony's independence was initiated by the early companies' attempts to minimize dependence on the unstable labor force. This was the basis for the establishment of the fifteen major sawmills. As regards the environment and the extraction of timber, by 1939, 20 percent or 5,832 square miles of the High Forest Zone was under reservation, but as noted by Barker et al. (1991, p. 23), "The colonial landlords had the foresight to establish reserves, but controlled the commercial timber harvesting for themselves and engaged in what amounts to timber mining in non-reserve areas." Britain, by the end of this phase, had been able to create a turf for the extraction of tropical timber, which ensured its control in the colony with respect to timber as a resource. It was, in fact, the Second World War, 1939, that initiated a more serious development of the wood-processing industry.

PHASE III: WORLD WAR II TO 1957

Ghana's tributary role as a British colony became more evident, as its timber and other resources were organized and siphoned off to support Western war efforts. During the war, the Forestry Department was run by a caretaker staff since the majority of the expatriate (colonial) officers had been drafted into the army. The Department organized the supply of forest products for use by the British and American Armies based in the country. This encouraged the production of lumber by pit-sawyers who produced as much as 1 million $ft.^3$ of lumber in 1944–1945 for the department, crude as the technology was. This was a precedent for private individuals, for purposes of survival, to use crude methods to legally or illegally produce lumber for the domestic market, which in certain cases involves encroaching on government turfs—the forest reserves.

Urban/Metropolitan Location of Firms

The domestic demand for lumber, coupled with the military requirements for the war increased the production of lumber. Apart from organizing pit-sawing, the Department put together derelict sawmill machinery and started lumber production in Kumasi, the capital of the Ashanti Region. Almost 50% of the lumber output was absorbed by the Armed Forces for use abroad. In fact, this sawmill, which became Wood Supply (Ghana Limited), alone produced a total of 83,277 cubic feet of lumber between August, 1942 and February, 1947 when it was sold. This contribution to Britain's war effort, based on derelict equipment and a strategic location had two effects in terms of the structure of the industry. First it served as a good precedent in terms of the pattern of dependence on pre-used obsolete technology in the industry, and second, it ossified the pattern of locating sawmills in urban centers, particularly in Kumasi prior to the independence era. Quite apart from establishing Kumasi as a major seedbed for subsequent location of new sawmills as a result of this sawmill's demonstration effects, it enhanced urban oriented location of sawmills. In other words, a model was established whereby rural locations where the log inputs are extracted—the traditional raw material-orientation of sawmills—became less preferable relative to urban center locations.

The selection of Kumasi, the capital of Ashanti Region, was important in terms of locational advantages for a number of reasons. It was the largest and most important urban center in the tropical high forest or closed forest zone from which the timber is extracted. Historically, it was the capital of the Ashanti Kingdom which was colonized by the British in 1902 and thus became the most important politico-administrative center and military outpost in the interior, next to the colonial headquarters, Accra, on the coast. It was the hub of the colony's transportation and communication network, the northern apex of the railroad network, and the most important center of commerce in the interior. Consequently, Kumasi had a concentration of people with relatively higher incomes who served as a major domestic market for the lumber and had a relatively more stable labor force including those who had migrated from the rural areas. Kumasi also had one of the best infrastructural facilities in the country at the time. The availability of such infrastructure was not only because of the economic and political power of the elite groups[4]—the administrators, intellectuals, and other professionals (see Richardson et al., 1986; and Lipton, 1982), in addition to the wealthy cocoa farmers and traders—but also as a colonial administrative center, as there was a concentration of expatriates in this location. These elitist expatriates' housing required the use of lumber, as their lifestyles required infrastructural facilities close to the level in their home countries. Also, it was easy to recruit both skilled and unskilled labor for the industry in Kumasi.

Furthermore, as a hub of the transportation network, the logs could easily be brought to Kumasi as a collection point from the surrounding forests for processing and then transported to the harbor (alongside the unprocessed logs) via the railroad for export. The railroad also made it easy to transport the imported processing machinery from the ports to Kumasi (see Mabogunje, 1973; Amin, 1974; and Slater, 1975). In sum, the pioneer sawmill in Kumasi, in the timber-rich Ashanti Region, served as a magnet for the location of new sawmills in that city and elsewhere through the process of cumulative upward causation, particularly via neighborhood and demonstration effects over time. In other words, urban centers with similar locational factors subsequently tended to attract sawmills thereby establishing the pattern of urban location of sawmills.

Two British-owned companies, African Timber and Plywood Limited (ATP) and Gliksten West African Limited (GWA), like Mim Timbers, however, did not locate in established urban centers. Rather, they initiated an alternative pattern for big companies by establishing township enclaves at Samreboi near Enchi, and Dwenase near Wiawso respectively, in the Western Region as the headquarters for their logging and sawmilling operations (see Fig. 2.1). This was because they were very rich firms and could afford to locate away from other firms to avoid competition and to be close to their huge concessions. Each of them secured concessions close to four thousand square miles. It thus became logical that they locate their processing activities on or close to those concessions. To make up for the deficiencies in urban infrastructure therefore, they established townships which were almost self-sufficient, in terms of the basic infrastructure required. These township enclaves had their own power generation systems fuelled by wood-waste, as well as their own hospitals, schools, water supplies, and residential accommodation for both higher and low management personnel. Skilled and unskilled labor and their families were provided with housing as well.

Production Technology, Equipment Sourcing, and Firm Ownership

We noted earlier that the beginning of sawmilling in Kumasi with derelict sawmill machinery by the government, which could have afforded the latest technology set a key precedent in terms of the pattern of dependence on pre-used, obsolete, and unsophisticated technology or equipment in the industry for the processing of timber into lumber prior to independence. This was reinforced by the fact that even the pit-sawyers with their relatively crude processing methods could produce large quantities of lumber for the Forestry Department. It was also the expatriates who had good knowledge of machinery used for wood-processing, especially, the Levantine and British traders and merchants. They also were the ones who had access to, or actually had the necessary capital to establish such facilities which required substantial financial outlays. The industry thus became traditionally, one dominated by expatriates using foreign capital since the first sawmill was established by the colonial officers and were followed subsequently by other expatriates, notably the British and the Levantines. Also the diverse origins of the expatriate investors established a tradition whereby the processing machinery and equipment were obtained from diverse sources. Consequently, the machinery and equipment tended to be of diverse makes and specifications. Between 1948 and 1950, another British firm, F. Hills and Sons Limited imported used American equipment, and installed a complete system comprising a sawmill, a "capital" slicer, a peeling and complete plywood line, and a flooring and flush door manufacturing facility, at Takoradi. This had two implications:

first, it initiated Takoradi as another seedbed with the benefits of the basic urban infrastructure and the port which made exports easy. Secondly, it further reinforced the pattern of using second-hand/used equipment and demonstrated that it was profitable to do so even with such equipment brought in from as far away as the United States.[5]

Lumber Exports and the Domestic Market

The export of lumber for the war effort also reinforced the colonial tradition of exporting a large proportion of the lumber output abroad. This was more so since generally, timber was not a popular and competitive material for domestic building construction (Combé, 1972, pp. 6–7). Furthermore, owing to the high cost of processing due to the industry's dependence on foreign requirements, external markets became indispensable as the main destinations for production to be cost-effective. This was also tied to the fact that consumers in the developed countries usually try to save, and to replace their own wood resources through massive timber imports from tropical regions since they work in forests where the rate of utilization is normally around 80% (Combé, 1972, p. 2). The tradition whereby little attention was paid to the domestic markets, with respect to the sale of lumber was reinforced and export-quality lumber became difficult and expensive to obtain on the domestic markets because of the priority given to exports.

World War II and the External Markets

The advent of WWII however, led to the disruption of global markets and currencies. Britain, for instance, was confronted with foreign exchange shortages and curtailed its purchases of timber from the United States. The war had led Ghana and other West African countries to emerge as a large source of hard-wood supply, booming the forest industry. The disrupted markets and the associated currency problems thus cut Britain and other large consuming countries in Western Europe off from their traditional sources, stimulating an unprecedented acceptance of West African timber and consequently, their replacement of U.S. hardwood in Western Europe. This intensified the pattern whereby the country's lumber is sent to the Western European market. It also initiated the same market-orientation for furniture parts and fixtures production which subsequently developed.

Capital Investments and Firm-Ownership Structure

As a result of the exports to Western Europe between 1946 and 1950, sawmilling expanded using almost entirely outside capital investment. Indigenous entrepreneurs could not raise such capital, hence intensifying the pattern of the industry's dependence on foreign capital. For one thing, they had not been able to accumulate the necessary capital to set up such outfits, and for another, the foreign-owned banks which were the only ones existing at that time, were not willing to grant them the necessary loans, a pattern that persisted through the period the ERP was adopted. As a result of the expansion of sawmilling with external capital, lumber exports increased as much as fourteen times in 1950 as exports in 1948. The domestic market which was now very small relative to the export market was of little consideration in the decisions to establish medium/large-scale sawmills. Nonetheless, it tended to reap some of the benefits, because all the high grade but non-export quality lumber as well as the rejects had to be disposed of on the domestic market.

Input Sourcing and Linkages

The entry of some major British companies into the timber industry in the late 1940s also initiated certain pre-independence patterns regarding input sourcing and linkages. Mim Timber Company Ltd., located at Mim, near Goaso in the Brong-Ahafo Region (see figure 1.1) was established in 1947 with the acquisition of large areas of forests as concessions to feed its sawmill using foreign capital. It did not see the need for integration until the seventies when it expanded to include kilns, a molding line, a subsidiary company—Scanstyle Mim Limited—to produce furniture and joinery-parts, as well as veneer-slicing facilities. Whilst the acquisition of large timber concessions by lease became an established pattern for the large sawmilling companies, they tended to avoid investing in integrated wood-processing industries. This saved them from the hassles associated with vertical linkages or downstream processing in a developing country, such as lack of skilled labor. Such processing, they figured, was more cost effective in their home countries in the sense that it could be done more easily and efficiently by subsidiary companies in their respective home countries while providing jobs for their home nationals. The avoidance of integrated industries also established a pattern of wastage or inefficient utilization of residual wood from the mills (e.g., sawdust, chips, and the medicinal barks of some of the species). In cases when these foreign companies chose to engage in integrated processing, it usually involved the production of plywood, veneer, and fixtures such as flush doors and floorings which involved the investment of huge sums of capital. As for the small and medium size wood-processing companies, they generally shied away from processing operations other than lumber production. They preferred the latter because it is quick income-generating and involved less capital investment.

This phase ends the colonial era, by the end of which most of the patterns such as a) production of lumber for Western European markets; b) location of firms in the urban/metropolitan areas; c) dependence on used imported machinery and equipment; d) domination of foreigners in the industry; and e) the use of capital-intensive methods of production, had been set. Subsequent phases—Phases IV-VI—involved attempts to undo or adapt the patterns, which had become largely problematic for national development. In the next chapter, we shall discuss the advent of Ghana's independence, and the structural changes that emerged in the timber sector and their implications for the country's development.

NOTES

1. *The Lagos Plan of Action for the Economic Development of Africa—1980–2000* (OAU, 1982) is used here as a standard development frame of reference, since the concept of "development" is a subject of very elastic interpretations. The LPA, as an action plan, included a detailed analysis and prescriptions in respect to such development issues as agriculture and industry. The plan was intended to help SSA countries achieve self-sustaining development by the year 2000.
2. Information regarding this was based primarily on Dickson (1969) and Peter Boateng's Report published in Barker et al., 1991.
3. It was in the last half of the nineteenth century that the British colonized the coast as the "Gold Coast." Later, they annexed Ashanti as well as the Northern Territories, all of which, excluding Togo (formerly, Trans Volta Togoland), make up modern Ghana.
4. See Richardson et al. (1986); and Lipton (1982).
5. F. Hills and Sons Limited was sold to a Ghanaian in the late sixties and renamed Ghana Primewood Products Limited.

Chapter Three

Independence, Instability, and Contested Turfs

Structural Changes in the Timber Industry

Ghana secured her political independence from Britain on March 6, 1957, under the leadership of Kwame Nkrumah. Politically, therefore, Britain apparently "lost" formal administrative control of an important economic and political turf on that day. This loss was however superficial because, as noted by Sawyerr (1990, p. 10), the country's social, political, and economic structures had already been re-ordered *via* the colonial process to integrate it into the world capitalist system as discussed in chapter 2. With the re-ordered political and economic structures already in place, an international capitalist turf had been created, and the timber industry, one of the dynamic sectors of the economy, had thus been aligned with the interests of the dominant classes of the metropolitan countries. In addition, the political and social conditions necessary to sustain and reproduce the relationships established under the colonial system had also been firmly established. Post-independence attempts by various governments to undo or adapt the patterns, which had become largely problematic for an authentic national development, were made by various governments like the Nkrumah and the first part of the Rawlings regimes but the stubborn resilience of capitalism, based on the established political and social conditions necessary to sustain and reproduce the relationship under the colonial system, eventually forcefully re-surfaced in the form of structural adjustment programs—a free-market-based economic policy—in the Neo-Liberal era.

Before examining the phases from the Independence era through the neo-liberal era, it is important to take an overview of the whole period as a starting point. From this vantage point it is apparent that overall, post-independence Ghana's governments have tended to sustain the international capitalist turf. This has been done mostly by strengthening the fundamentals of its extractive economy, of which the timber industry constitutes a major part, particularly for the transnational corporate sector, as the state continues to ensure cheap access to the forest communities' assets on their behalf as shown in the FERN Report (FERN, 2006). In so doing, the Report notes, the state continues to abuse the communities' rights, especially as the international commodity markets continue to undervalue natural resources and transfer wealth to international capital, sometimes exemplified by food imports as part of the associated domestic economic disruption. In specific relation to the timber industry, the Report argues, the transfer of the power of patronage to Ghanaian politicians by the British changed the internal composition of the timber industry and its relationship with the political and forestry establishments, as reflected in hundreds of indigenous supporters of the successive civilian and military regimes of Ghana, who joined the ranks of the domestic and international capital—the logging concessionaires—in putting additional stress on the timber resources. The activities of these supporters underscore the view that the formal and informal sectors constitute turfs within which capitalist and pre-capitalist elements struggle for control as shown in chapter 1. Since the state in essence, is "suspended" to a degree over the informal sector (see Hyden, 1980), people, including some government officials, party or government supporters, are able to move in and out of both sectors to operate whenever deemed necessary, convenient, or personally profitable (see Sarfo-Mensah, 2005).

More importantly, from a governance perspective, the successive civilian and military regimes changed the character of "regulation" completely. This is because, as the FERN Report (2006) notes, while international capital—the British concessionaires—were largely well-established corporations (such as Unilever) employing professional foresters and lawyers to manage their relationships with the state, few of the new timber contractors—often with ties to the government—understood or were interested in a rules-based regulatory system, and none worried about environmental impact or sustainability issues. These individuals were politically savvy and willing to use their connections to override regulatory restrictions on their activities. Thus, it was (and remains) difficult for underpaid forestry officials to assert authority over wealthy and politically influential "Timbermen," especially where these proved willing to offer substantial rewards to "flexible" officials. As their wealth grew, "timber contractors" freed themselves from their original political sponsors, spread their risks, and became patrons of the national political elite as a whole. In time, they extended their influence from politicians to more powerful silviocrats and throughout the forestry system to its lowest operating levels. Normally, though the superstructure serves the dominant interest, it does so with some independence and in accordance with the internal logic of, for example, the legal system. However, as noted in the Report, the tendency to provide instant gratification to powerful "Timbermen" ultimately undermined the integrity of the superstructure. Between 1948 and 1992 there was no major policy review despite political independence, "Africanization," international commodity price collapse, nationalization, and the many shifts in the political-economic function of the forest sector. Politicians and civil servants ran the sector by *ad hoc* administrative fiat; governments passed laws to catch up with administrative pronouncements, or not at all. Management plans mostly expired in the late 1960s and were not replaced. With the increasing power of the timber industry, the Report further shows, sector governance capacity atrophied, and by 2004, the situation had degenerated to the point where the Forestry Commission Board could unashamedly justify flagrant abuse of the statutory permits regime with the claim that it had acted in "the best interest of the timber industry and the nation" (FERN, 2006, pp. 20–21).

In relation to the above, it can therefore be said that developments in the timber industry with respect to turf and trade began to significantly change during Phase IV, dominated by the socialist orientation of the Nkrumah administration, which was toppled in a military coup d'état in February 1966.

PHASE IV: 1957–1966

By the time of Ghana's independence in 1957, then, the basic structure of the industry had been well established with various implications for economic development in Ghana. Regarding the timber trade, for example, the largest foreign-owned sawmills had assured markets abroad since their parent companies in Europe undertook the marketing of the lumber themselves, or else utilized lumber in the manufacture of timber products. The majority of the firms however did not have overseas agents. They therefore marketed their lumber through exporting companies. Contracts for exports were generally obtained before production began, and in slack periods these sawmills had to produce in anticipation of demand. Such firms thus tended to produce under uncertainties during such periods. The only assistance the government provided in that respect was to provide rail space to ease some of the financial strain on some of the companies through the Timber Marketing Corporation established in 1957 (Chryssides, 1974, pp. 67–68).

The big European timber companies also established their own turfs by securing large timber concessions. Access to such extensive concessions as those of Mim Timber Company, African Timber Products Limited (ATP), and Gliksten West African Limited (GWA) was possible because until this phase—specifically, in the 1960s, when two landmark legislations—the Administration of Lands Act and the Concessions Act were enacted—the traditional "Stools" and other tribal groups possessing their own forests could allocate their concessions directly to contractors and companies. This facilitated the use of money, power, and influence by the big companies to secure concessions, hence there arose the need to centralize the whole allocation process as a result of administrative difficulties and the litigations arising out of the fuzzy nature of the concessions' boundaries. The promulgation of the Administration of Lands Act of 1962 (Act 123) vested the stool lands in the Head of State, making him/her the trustee of all such lands. It also permitted a standardization of the limits relating the duration and size of a concession. A grant of any timber rights was not to exceed 30 years in duration and a single grant was not to exceed 40 square miles, and an aggregate of grants of timber rights was limited to 240 square miles.

The Concessions Act of 1962 (Act 124) was used to replace the colonial administration's Concession Ordinance of 1900. This Concessions Act empowered the Head of State to cancel concessions if they were deemed to be "prejudicial to public safety or interests" and was mainly directed at excluding expatriates; however, it also inadvertently promoted corruption in the industry by providing opportunities for expatriate firms to collude with willing Ghanaians as fronts, to keep their interests. This raises the issue of firm ownership in the timber industry and the issue of corruption.

Firm-Ownership Structure and Corruption

In the 1960s also, foreign-ownership of companies entered a new phase, with the entry of Levantine families—particularly Lebanese and Syrians—into the timber sector. Unlike the British, the Levantines established their mills from capital generated from retail trade in automobile spare parts, pharmaceuticals and clothing, and significant support from the local foreign-owned banks. Support for foreigners from the local foreign-owned banks was significant because one of the pre-Independence patterns was that those banks, like Barclays, discriminated against indigenous Ghanaians in the granting of credit facilities (Quarcoo, 1990, pp. 9–10; World Bank, 1991, p. 83).

The Levantines, who felt that they were being discriminated against by the Forestry Division, because they found it hard to secure concessions for their operations as a result of the Concessions Act, 1962, tended to locate their mills in the established large urban centers of Kumasi and Oda in the Ashanti and Eastern regions respectively. Their business practices also initiated a pattern of corruption and fraud, particularly with respect to the transfer of money overseas as they sought to maximize and secure their profits. To do this, they often tended to misrepresent their timber exports in terms of species, volumes, values, and destinations (Chryssides, 1974, p. 65).[1]

In fact, this was even exacerbated in the next phase because many of the Lebanese and Syrians who had been in Ghana for generations felt that their future in Ghana was so uncertain due to the passage of the Ghana Business Promotion Act (Act 334), 1970, which excluded aliens from certain types of business activity, on top of the exclusionary Concessions Act, 1962. Anticipating that such exclusionary legislation may be extended to smaller sawmilling enterprises, they shelved plans for expansion for day to day operations and sought ways which were not always legal, to maximize their income and transfer most of it offshore. It was therefore the attempts to maximize their incomes under the circumstances and ensure that such monies were safely repatriated that established the pattern of illegal transactions, fraud, and the corruption of public officials (Chryssides, 1974, p. 65). The public officials involved tended to have a strong incentive to demand rewards knowing how much of a premium the expatriate beneficiaries get from the exercise of their administrative power (see Botchwey, 1993, p. 26).

On March 5, 1957, the day before independence Kwame Nkrumah stated his intention to reclaim part of the capitalist turf. He declared to Parliament that "quite clearly, one of the first things which must follow independence is that citizens must play a far bigger role in the commercial and industrial life of the country than they do at present" (see Adu Boahen, 2000, p. 197). As part of his socialist industrial development program, Nkrumah's Convention People's Party (CPP) administration established the State Furniture and Joinery Corporation in Kumasi to produce furniture and fixtures for government institutions and organizations such as the universities, hospitals, schools, and government offices. It was therefore during this phase that formal sector production of furniture entered the scene. Hitherto, the production of furniture and fixtures had been carried out on very small scales such as carpentry shops, mostly in the informal sector, and as attachments to some building construction outfits largely to meet the needs of the domestic market. In fact, prior to 1960, the marketing of round and processed-wood differed according to the size and ownership of firms, and this leads us to a discussion of the organization of the industry during this phase and how the processed and unprocessed wood were marketed.

Marketing and the Organization of the Industry

The Timber Marketing Corporation, created in 1957, developed into the Ghana Timber Marketing Board (GTMB) in 1960 with the responsibility of organizing the timber industry and providing it with the necessary assistance. In 1963, it was charged to fix, periodically, the minimum prices of all logs and sawn timber, open the trade to all producers and resident (expatriate) buyers, and free them to make their own arrangements regarding the purchase, transportation, export, or domestic sales of wood and wood products. In essence therefore, it was a mechanism whereby the socialist government was trying to take over and control a marketing turf that hitherto had been established and controlled by foreign capital in its accumulation process.

The bureaucracy associated with the GTMB, as the sole government agency charged with the overall organization of the industry, and its monopsonistic status established a pattern of inefficiency, corruption, and mismanagement that affected the overall organization and operations of the industry (see Killick, 1978). The activities of the GTMB thus became inimical to the growth of the timber industry, in particular, and the nation's development in general, a deterioration that festered during Phase V and reaching its peak by 1983 when the free-market based Economic Recovery Program was adopted.

PHASE V: 1966–1983

February 1966 marked the end of the socialist administration of the Nkrumah government through a military coup d'état. In that year, the new government of the National Liberation Council which toppled the Nkrumah regime, sought to promote indigenization or "Ghanaianization" of business ownerships by indicating that "the government is determined to promote as rapidly as possible the development of an efficient Ghanaian entrepreneurship capable of running any type of modern business in all sectors of the economy (Ministry of Information, 1968, 1).[2] This facilitated the entry of Ghanaian logging companies into wood-processing in the Seventies. Prior to that, there was only one, Bibiani Logging and Lumber Company Limited, which had been established in 1949 as a logging concern. It then added sawmilling to its operations in 1960. The entry of the Ghanaian companies was facilitated by the creation of a suitable business environment for them by the government over the period between 1972 and 1978. Over that period, the National Redemption Council (NRC)/Supreme Military Council (SMC) regime sought to capture "the commanding heights of the economy" from foreign domination by nationalizing important industries or taking majority shares—in other words, capture and control a turf that had long been created, held, and controlled by foreign capital. Specifically, the NRC government issued the Investment Policy Decree in April 1975 which *inter alia,* required full Ghanaian ownership of wholesale trading-firms and partial ownership of all other firms. In connection with that, the government for example, acquired 55% of the shares in major wood-processing companies as Mim Timbers Limited and Gliksten West African Limited. A few years later, the foreign minority shareholders sold out to the government making the latter the sole owner of such companies. The takeovers were thus not preceded by formal feasibility studies and project planning; hence such companies were soon plagued by "shortages of critical staff, overstaffing, inadequate capitalization, undue political interference, mismanagement, corruption and inefficiency" (Gyimah-Boadi, 1991, p. 195).[3] As a result, the government began to overextend itself in terms of financial resources because such industries became a drain on national budget resources (Killick, 1978). Thus, a pattern that emerged here was that such public-owned companies were characterized by poor management (dominated by unqualified personnel who were mainly political appointees), lack of finance, a shortage of spare parts, and other operational problems like dwindling labor productivity and very low levels of capacity utilization. Government ownership of the major firms therefore tended to seriously undermine the industry.

Overall, as a consequence of the historical development of the industry from Phase I through Phase V, five main problematic structural and organizational patterns were clearly evident in the industry by 1983, when international capital forcefully reclaimed its turf via the IMF-World Bank-led structural adjustment program, dubbed Economic Recovery Program. The patterns related to: 1. Production and Marketing; 2. Location and Linkages; 3. Equipment, Technology, and Employment; 4. Foreign Exchange Generation; and 5. Firm-Ownership and Organizational Structure of the Industry. These problematic patterns had persisted right through a whole series of very different types of governments since the beginning of Phase IV, in terms of not only internal ideological persuasions or orientations, but also their attempted strategies *vis à vis* international capital, and other interests such as private domestic capital, and resident expatriates (especially, the Lebanese and the Syrians). It therefore did not seem to matter very much which strategy was attempted, for the patterns never really changed all that much for the better. The respective features the patterns are discussed below as they all have implications for spatial integration.

THE PROBLEMATIC STRUCTURAL AND ORGANIZATIONAL PATTERNS IN THE INDUSTRY BY 1983

Production and Marketing

By April 1983 when the ERP was adopted, one of the major characteristics of the timber industry was that the bulk of the timber harvested was exported in the form of logs, as shown in figure 3.1. Processed wood exports were

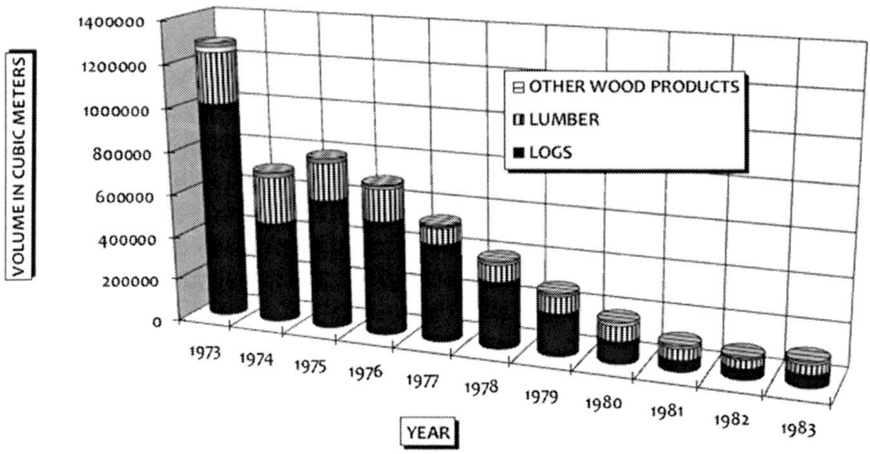

Figure 3.1. Volume of Timber Output Exported to Europe (1973–1983)

significantly low. Log exports in 1973, for example, constituted as much as 78% of the total exports to Europe as compared with 19% for lumber and 3% for other wood products. In terms of volume, it was 100,6848, 234,829, and 41,049 cubic meters respectively. Even in 1982, when the industry was at its lowest ebb, the proportions were 53%, 40%, and 7% for logs, lumber, and other wood-products respectively. This shows that in terms of actual processing of the timber, the industry was dominated by sawmilling, lumber from which was mainly exported to Western Europe particularly Britain.

That had been the pattern after independence, as noted earlier and as borne out by figure 3.2, which shows the proportions of lumber output exported in the Sixties, indicating that at least 51% of the annual output was exported. In 1970, whilst recorded exports amounted to 240.8 thousand m^3, domestic sales were only 94.1 thousand m^3. In 1978, five years before the ERP, recorded exports and domestic sales were 170.0 thousand m^3 and 160.0 thousand m^3 respectively. From 1981 through 1983 however, when the industry was at its lowest ebb, domestic sales of the lumber exceeded exports. Whilst domestic sales registered 100.0 thousand m^3, 90.0 thousand m^3, and 90.0 thousand m^3 for 1981, 1982, and 1983 respectively, the corresponding figures for exports were 50.0 thousand m^3, 40.0 thousand m^3, and 50.0 thousand m^3 respectively (World Bank, 1991, 117). There was thus a shortfall in lumber supply relative to the demand in Western Europe.

In comparison with other downstream mill processing, over the period 1978 through 1983, the highest veneer output was 36.6 thousand m^3 in 1978, and the highest plywood output was 63.2 thousand m^3 in 1981. In the case of veneer, exports always exceeded domestic sales whilst the opposite was

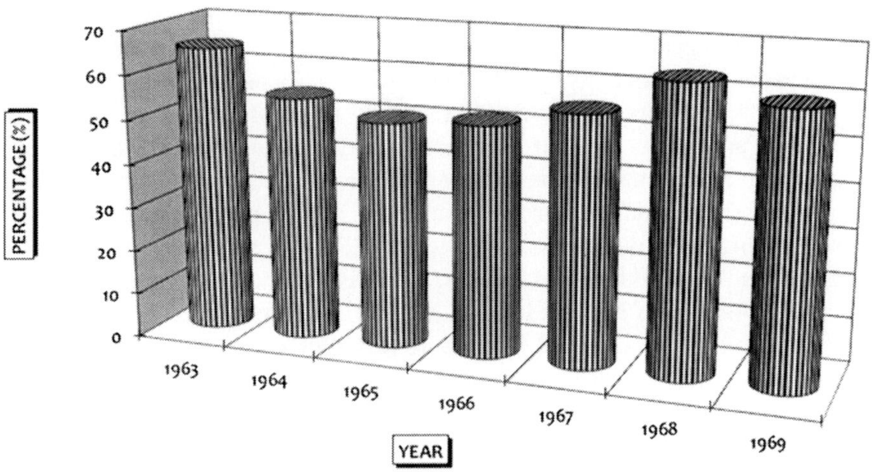

Figure 3.2. Proportions of Lumber Output Exported to Europe (1963–1969)

true for plywood over the period 1977–1983 (World Bank, 1991, p. 117). Relatively less emphasis was therefore placed on secondary stage processing or the production of such tertiary items as furniture and fixtures. The predominance of log exports vis à vis processed wood exports by the timber companies was due to the minimal production and export processes involved in the export of logs. All the processing costs (e.g., equipment, specialized labor, and transportation) associated with lumber and other processed wood products are thus avoided whilst huge profits are still made (Boateng, 1991, pp. 30–31). The fact that the bulk of the valorized components of the timber exports were dominated by lumber, indicates that tertiary processing was very limited even though the foreign exchange contribution generated by processed wood was very substantial. Of the total export revenues generated by the timber sector in 1973 (US$ 136,172,718), 29% (US$ 40,148,792) was contributed by lumber, and 8% (US$ 11,474,876) by other processed-wood products. In 1982, they were 24% (US$ 8,973,462) and 17% (US$ 2,621,475) out of the total US$ 15,295,728. Weather and road conditions often determined the availability of logs for processing and thus affected the volume of output. Many sawmills therefore had to establish a log stock sufficient for one month, at least, to maintain the production for export (Combé, 1972, p. 61).

It was the establishment of the Timber Marketing Board in the 1960s that gave an impetus to the growth of downstream production. According to Combé (1972, p.70), the government's decision to control the timber trade by making the Timber Marketing Board a monopsonist institution led to a very sharp drop in log production and export and depressed the industry. The lack of export opportunities for logs at the beginning of the 1960s therefore led to the emergence of a number of small wood-processing establishments which were stimulated by massive government loans to revitalize the industry. However, since these new processing firms required foreign capital equipment and inputs for their operations, they became externally dependent. Such dependence entailed consumption of part of the country's very limited stock of foreign exchange. As shown in figure 3.3, the percentage of imported inputs to total material inputs, prior to the ERP, for saw/wood milling and planing,[4] and for furniture and fixtures, were very high. The former consumed as much as 43.4% of imported inputs in 1977, and the latter, 42.9% in 1982.

Production in the industry for export was thus characterized by a small degree of valorization, in terms of manufacturing value added. This was more so since the lumber was exported in the air-dried form. Drying the lumber via the kiln-drying process before export would have increased its value by 15%–25%. Unlike the air-dried lumber with its varying moisture content levels, kiln-drying pegs the moisture content level at the specification of the foreign importer. Kiln-drying is therefore a prerequisite for processed-wood export items as moldings and furniture parts to avoid wilts and warps which alter design specifications.

The small degree of valorization in the industry was also in part, the result of lack of processing equipment and skilled labor. These were in addition to the other reasons adduced to the prevalence of timber exports in the log form. Processing the wood as a measure for export valorization was tied into the patterns relating to the location of the processing firms and the associated linkages.

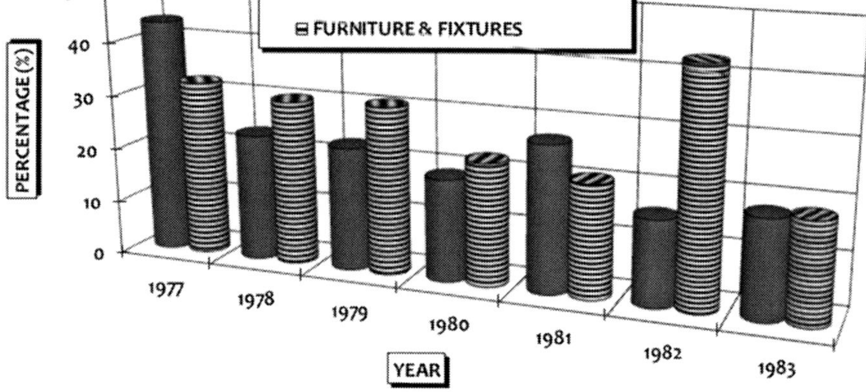

Figure 3.3. Percentage of Imported Inputs to Total Material Inputs—Processed Wood

Location and Linkages

In terms of locational characteristics, the wood-processing firms were mostly concentrated in urban or metropolitan centers located in the various administrative regions within the Closed Forest Zone, as noted in chapter 2, due to the historical development of the industry. This locational pattern of the firms must however be linked to their sources of inputs and their channels of distribution, which raises the issue of the nature of the roads, railways, and the ports, as well as telecommunications prior to the adoption of the program. By 1983, these basic infrastructures had largely been neglected. For a long time, the roads and railways had been poorly maintained and most sections were in a very bad state of disrepair. The ports had also not been upgraded over a long period and so had become obsolete and very inefficient in terms of operations and facilities available (e.g., there were no container-handling facilities). The telecommunications system had also been neglected and had become very obsolete and unreliable. All these were mainly the result of economic mismanagement (Huq, 1989; Rado, 1985). This was particularly so during the administrations of the Supreme Military Council I, under Colonel I.K. Acheampong (January, 1972 to July, 1978) and the People's National Party, under Hilla Limann (September, 1979 to December, 1981). The neglect thus greatly increased the cost of business operations by the firms (World Bank, 1989, p. 5) which translated into extra costs for firms based on their respective geographical locations.

Ghana currently has 10 Administrative regions (see figure 3.4). The Closed Forest Zone which contains the tropical timber mainly covers the Western Region, Ashanti Region, part of the Brong-Ahafo Region, Eastern Region, and the Central Region. In terms of regional location of firms and production, Chryssides (1974, p. 62) documents that there were 62 sawmills in production in Ghana in 1969. The Ashanti Region with 33 mills—most of which were in Kumasi, the original seedbed—was the largest producer of sawn timber and accounted for more than half of the domestic timber sales. The sawmills in the Western Region, also in urban centers like Sekondi-Takoradi and Samreboi, were mostly larger in size, with five mills in this region producing over 30% of the nation's total lumber output. The Western Region also served as the largest exporter of lumber. All the five sawmills exported the greater part of their lumber production.

Sawmills in the Eastern Region, on the other hand, were mainly domestic market-oriented and their average size of sawmills was smaller than in any other region. In general, locally oriented sawmills tended to be smaller than the export-oriented ones. The sawmills in the Eastern Region, also based in urban centers like Oda, Koforidua, and Nkawkaw, supplied lumber mainly to building contractors and furniture producers in the Accra-Tema area. The

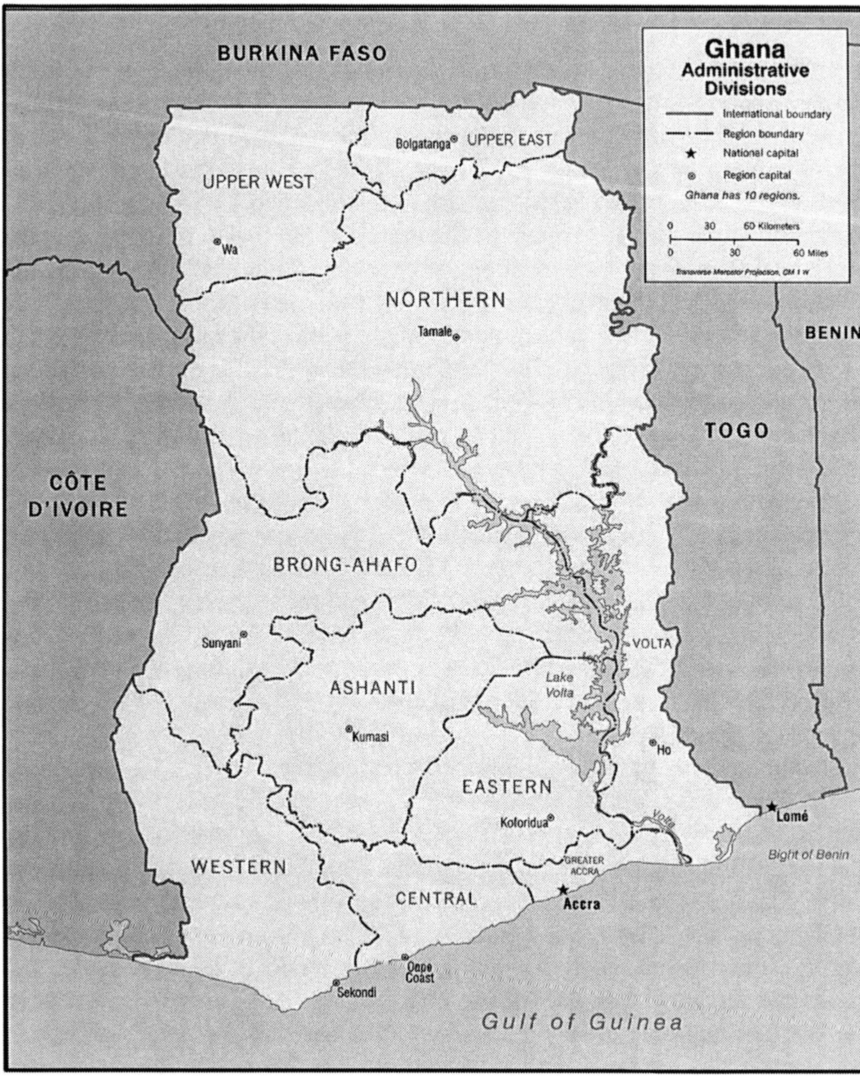

Figure 3.4. Regional (Administrative) Map of Ghana

production of furniture and fixtures was however carried out mainly in the urban centers where the markets existed. The 1979 issue of the *Directory of Approved Industrial Enterprises* of the Ministry of Industries, Science and Technology documented a total of 248 wood-processing industries. Of these, 77 were located in the Greater Accra Region (a metropolitan area), 81 in Ashanti Region, and 33 in the Eastern Region (both of which are relatively highly urbanized and have a high incidence of urban centers). The regions with relatively very few urban centers—Brong-Ahafo, Central, Northern, Western, and Volta—had 9, 17, 4, 26, and 1 respectively.

The regional distribution and urban concentration of the wood-processing firms reflect the polarized development in the country prior to the ERP. There were regional disparities as shown by the regional distribution of the firms; and rural-urban disparities as indicated by the concentration of the firms in urban centers. These disparities had been a consequence of circular and cumulative processes, namely a spiral build-up of certain areas as a result of some initial advantages, achieved at the expense of other areas (Myrdal, 1957), and the urban bias of the time, whereby policy makers paid more attention to urban areas to feather their own nests (Lipton, 1982) through the course of the country's history. It must be noted that to a degree, the disparities in the development patterns and investment allocations are "an inevitable condition of growth itself" because all areas cannot develop at the same pace simultaneously (Hirschman, 1958). In fact, policies based on equity or political exigencies to eliminate regional differences are often not the most suitable solution to the problem (Alonso, 1968), and may be rather very expensive. However, the concomitant urban problems such as the migration of rural people to urban slums (Busia, 1953; Caldwell, 1969; Lewis, 1967), the high cost of city maintenance, urban unemployment, congestion, and pollution, necessitate investment policies and choices to narrow rather than widen the regional differences (Darkoh, 1973).

The problem of polarized development was also associated with weak sectoral and inter-regional linkages (Szereszewki, 1966; Darkoh, 1971, 1973). As noted by Darkoh,

> the dominating structural feature of the economy of Ghana was the relative unimportance of flows inside the productive system. The sectors have little dependence upon each other; most of their output goes directly to final uses. Ghanaian manufacturing in 1960 supplied only 25% of its outputs as inputs to other sectors of production and 75% satisfied directly final demand (Darkoh, 1973, pp.10–11).

In addition, he also noted that the processing firms downstream were:

> seldom integrated in the national economy in a manner calculated to set off chains of growth in any significant direction. The utilization of by-products and waste material . . . (was not) a sizeable industrial activity; backward and forward linkages . . . only a limited role . . . and (had) been confined to mainly to simple backward linkages. . . . This (could) be seen in the scale of dependence on imports for intermediate and capital goods (Darkoh, 1973, p. 9).

The only significant intra-sectoral linkage, as observed by Combé (1972), was the forward linkage between sawmilling and the manufacture of furniture and fixtures. A notable consequence of the lack of intra-sectoral linkages was the problem of wastage.[5] Wood recovery rates were very low averaging 42% (Chryssides, 1974). The Forestry Commission (1988, p. v) acknowledged the problem by pointing out that "[t]he subsector needs more efficient wood-conversion systems, if it should minimize waste while maximizing on rates of recovery during the processing of log and other raw material base. This should be the only way to justify more domestic processing . . ." The problem of waste associated with the very limited intra-sectoral linkages was reflected in the fact that within the industry, the highest yields were obtained in mills which process logs into lumber for their own subsequent use. In other words, the more integrated the firms were, the less wood was wasted. Smaller pieces of wood or off-cuts were converted into such items as crates, moldings, and floorings. In most firms however, the off-cuts were disposed of as fuel wood, and the massive quantities of saw dust generated had no use at all, rather they constituted fire hazards. Regarding inter-sectoral linkages, Combé noted:

> the building industry also benefits greatly from the local production of sawn wood; it uses about 60 per cent of the locally produced lumber mainly for door and window frames, doors and structural work. On the other hand, the dependence of . . . sawmillers on the local manufacture of metal products, machinery and transport equipment also has a stimulating effect on the economy, although it is impossible to establish the exact relationship in this area. More accurate information is available regarding the freight train traffic, where timber . . . was the major commodity conveyed during the 1960s . . . Thus within the intersectoral trade, timber transport . . . greatly contributed to the total freight traffic of the Ghana Railway. With regard to locally produced electricity, the Ghanaian woodworking industries consume a minimal amount, despite the great number of machines running in the numerous enterprises. This is partly due to the decentralized locations of these industries[6] (Combé, 1972, pp. 111–112).

With respect to linkages therefore, inter-sectoral linkages were more developed than intra-sectoral linkages. The location pattern and the nature of linkages in the industry are also connected to the patterns relating to technology, machinery, equipment, and employment.

Technology, Production Equipment, and Employment

The structure of the timber industry prior to the neo-liberal period was generally characterized by a heavy dependence on capital-intensive, albeit outmoded technology vis-à-vis labor-intensiveness in the face of the country's cheap abundant labor. Quite apart from the historical basis of this pattern, wood processing particularly for the export market, requires specialized machinery based on modern technology, which are often protected by patent laws (Combé, 1972, p.116). Consequently, such technology tended to be avoided by the wood-processing firms because it consumes considerable amounts of foreign exchange.

The structural problem with this pattern is seen in the fact that after all, with the crude technology of pit-sawing, which was labor-intensive, as much as one million cubic feet of lumber was produced for the Forestry Department within the short period of 1944–1945. Such heavy dependence on capital-intensive technology thus minimized the industry's employment potential and thus denied employment to the numerous unemployed in the economy. In the case of sawmills, the size of the enterprise generally determined the degree of mechanization; and it was only the small-scale sawmills that were relatively labor-intensive (Combé, 1972, p. 54). The problem was thus with the medium and large-scale sawmills which were predominant. In the case of furniture production, for example, the industry had the option of producing unique furniture items, using labor-intensive methods; for example, the production of doors and panels uniquely engraved with Ghanaian symbols or designs were very successful. Even apprenticeship in the production of such unique items constituted needed employment of unskilled labor. The argument here is not that modern technology was to be dispensed with, but that the mix needed to be dependent on the nation's resource endowments so as to minimize unemployment and to make the nation as self-reliant as practicable (Seers, 1969, 1977; Mabogunje, 1981). For instance, The United Nations Industrial Development Organization (UNIDO) has pointed out that there is the need to enhance firms' capacities to adopt imported technology on the one hand, and to create appropriate indigenous technologies suited to local resource endowments, on the other (see UNIDO, 1986, pp. 72–73). UNIDO emphasizes strategies whereby industries develop or adapt the type of technologies which are appropriate to human and material resources of the economy. The strategy does not mean a mere transfer of modern technological processes available, but the adoption of the type of technology that will ensure optimum allocation of national resources.

In the face of Ghana's cheap abundant labor, employment opportunities and job creation are critical to the growth of the economy and development in general. According to the 1970 population census, Ghana's population was 8,559,313. By 1984, around the time the country adopted its structural adjustment program, the number had jumped to 12,296,081 (Statistical Service, 1989a, p. 93). Within this inter census period, the manufacturing industry as a whole employed a total of 54.0 thousand people in 1977 (Central Bureau of Statistics, 1983b, p. 43), this had decreased to 22.6 thousand and 43.3 thousand in 1982 and 1983 respectively (Statistical Service, 1991, p. 47).

For the wood processing sector, the period 1977 through 1983 witnessed a decreasing trend in employment with 18.3 thousand employees engaged in sawmilling, planing, and other woodmills, and 2.6 thousand employees in the furniture and fixtures sector. Just before the adoption of the structural adjustment program in 1983, these figures had dropped to 13.3 thousand and 1.2 thousand respectively. Figures 3.5a and 3.5b show that even during the hard times the industry was going through over the period 1977 through 1983, the net values of the fixed assets for both sawmilling and the production of furniture and fixtures rose relative to emoluments paid to employees. This is particularly significant since it was employment that suffered over the period.

As shown in figure 3.6, the number of operatives steadily fell from 14,651 in 1977 to 11,404 at the end of 1982 for sawmills, planing, and other woodmills. For the production of furniture and fixtures, there was also a steady fall from 2,185 in 1977 to 891 at the end of 1982 (Central Bureau of Statistics, July, 1983; Statistical Service, December, 1989a).

The relative rate of decrease was far less in the case of professionals or skilled labor employed (see figure 3.7) to support the fixed assets or installed capital. The drastic fall in employment in the production of furniture and fixtures strongly reflects the preference for capital-intensive production methods especially since the substitution of labor for capital is much easier in that sector. Darkoh (1973, p. 12) also observed that in general, capital outlay per worker among firms producing furniture and fixtures was relatively one of the lowest in all manufacturing in Ghana. This was however the reflection of the inclusion of the numerous informal sector carpenters who used very simple tools. According to Cantor (1967, p. 108), the industry tended to be relatively capital-intensive

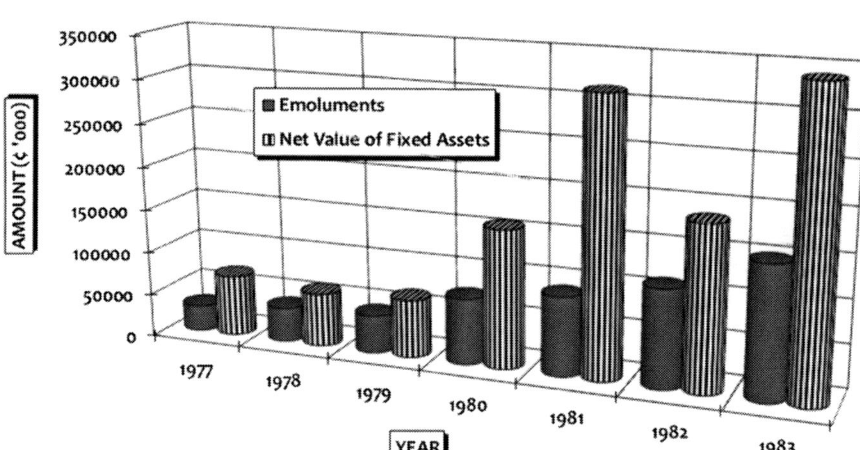

Figure 3.5a. Emoluments and Net Value of Fixed Assets for Sawmills (1977–1983)

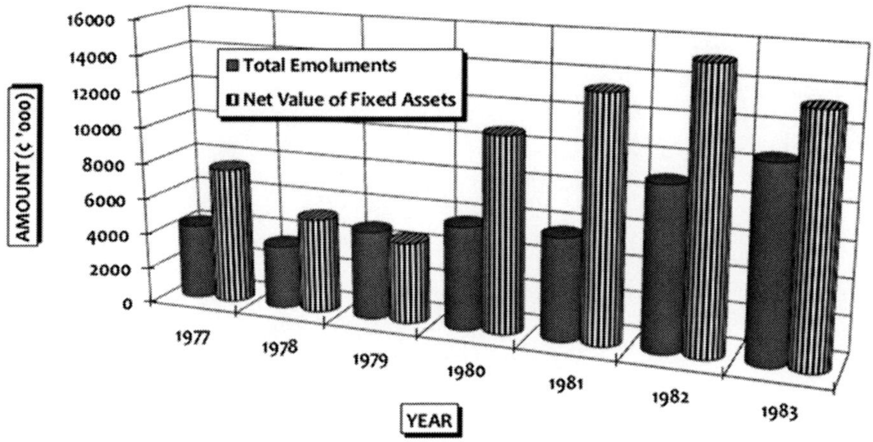

Figure 3.5b. Emoluments and Net Value of Fixed Assets for Furniture & Fixtures (1977–1983)

owing to the scarcity of skilled labor. Also, since the construction of modern industries required the importation of expensive capital equipment and machinery and often in addition, semi-processed materials and parts, it increased the capital to labor ratio. It must however be noted that with some of the imported technology, special skills are indispensable and therefore require expatriates who supervise the operations and thus guarantee essential repair and maintenance. This minimizes down time and ensures that the equipment stays in use for a reasonable length of time (see Boateng, 1991, p. 30).

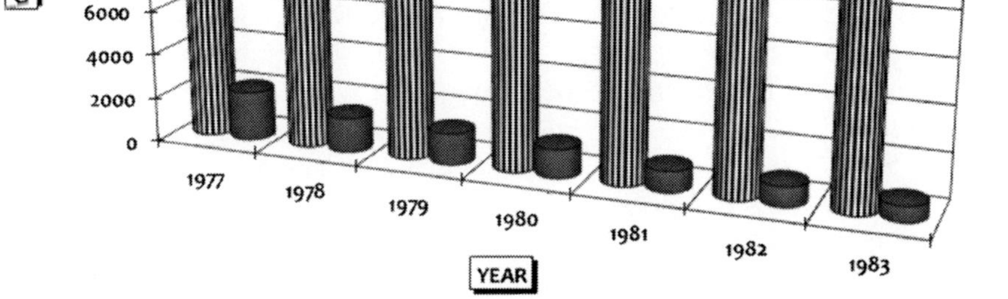

Figure 3.6. Operatives in Sawmills and Furniture & Fixtures Production (1977–1983)

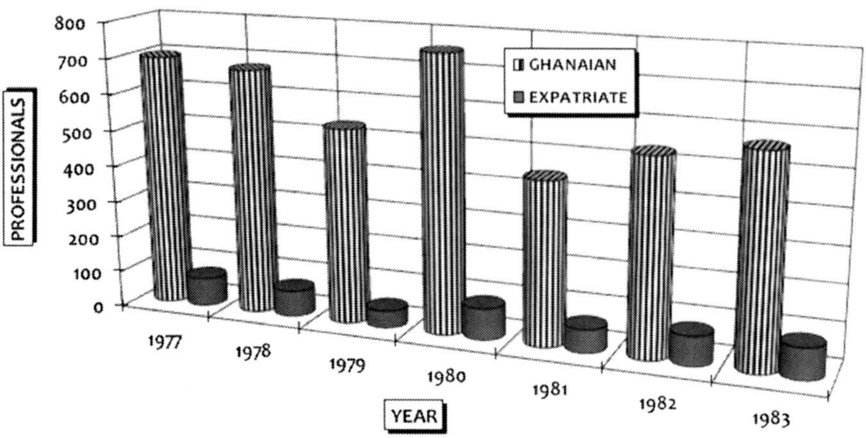

Figure 3.7. Professionals in Sawmills and Furniture & Fixtures Production (1977–1983)

In general, production and productivity in the industry were very inefficient. The production of furniture and fixtures was particularly characterized by mediocre-quality finished products. Combé (1972, p. 52) suggests that these were mainly due to lack of specialists and the inadequacy of the processing equipment. Capital-intensive technology thus became the favored option to overcome that problem. In addition, precision, quality control, and market requirements of foreign buyers of wood products usually called for the employment of foreign (capital-intensive) technological know-how (Boateng, 1991, p. 30). Capital-intensiveness therefore continued despite the readily available trainable cheap labor with a little more investment.

Regarding the mediocre quality of the products (which accounted for the almost non-existent export of furniture, furniture parts, and fixtures prior to 1983, Scanstyle Mim Ltd. being the only exception), particularly, with respect to specifications set by overseas customers, Combé (1972, p. 117) notes that the situation was reinforced during the period, because the number of Ghanaian technicians was very insufficient, coupled with limited training possibilities. According to Fieldhouse, there were very few schools or technical colleges to provide industrial training and managerial skills; hence by the end of the colonial era, *there was no stock of men trained in the skills of western business and that the tiny elites . . . (were) anxious to fill the more lucrative positions in government or education but rarely equipped or motivated to engage in productive enterprises* (Fieldhouse, 1986, p. 35). The shortage of technical and managerial skills was also exacerbated in the pre-ERP period by the immigration law that sought to promote indigenization—the "*Ghanaianization*" of all professions, as a measure to assert control over the production turf that had largely been dominated and controlled by foreigners. The law (Ghana Business Promotion Act 334, 1970) established quotas with respect to employment in certain occupations—limiting the number of non-Ghanaians who could be employed in those occupations. This therefore limited the number of expatriates who could be employed by various wood-processing enterprises. No immigrant was therefore permitted to accept gainful employment unless their occupation was among the authorized quota. This therefore contributed to the dependence on imported capital-intensive technology as a substitute for the unavailable or curtailed skills. The reliance on imports of technology, machinery, and equipment thus tended to mesh the Ghanaian economy into overseas marketing complexes thereby heightening the country's external dependency (Siedman, 1971). This had several negative implications not only for the industry, but also for the Ghanaian economy in general. According to a United Nations study (United Nations, 1970), the high capital-labor ratio derived from high installation costs, heavy reliance on expatriate skills, high transport costs, especially of inland locations, absence of organized commercial machinery within the country, delays and inadvertence, provision of high social overheads (e.g., expatriate housing), excessive use of suppliers' credit, ignorance at the purchasing end, and deliberate malpractices. In fact, it was in connection with such deliberate malpractices that the National Redemption Council military government of Colonel I. K. Acheampong unequivocally repudiated US$ 94 million of debts owed to some British firms and cancelled the 180–day credit system under which some short-term debts accumulated (see *Legon Observer*, Vol. VII, Feb. 11–24, 1972, pp. 60–66; and *West Africa*, week ending 18th Feb., 1972, p. 175 and week ending 25th Feb., p. 215).

The UN study also noted that the trend tended to be reinforced by the fact that the expatriate entrepreneurs, managers, and technicians, who usually came from advanced countries, based their investment choices on their own immediate experience and awareness of available machines and desirable levels of mechanization (UN, 1970, p. 9). This tended to make the pattern of capital-intensiveness in the industry, very expensive. Another reflection of the expensive nature of the external dependence, a legacy of the colonial era, was the nature of the imported capital equipment in terms of efficiency and cost effectiveness. In most cases, the machines were obsolete pre-used ones (Combé, 1972; Chryssides, 1974). Also, there was "an extraordinary diversity among machines in use" (Combé, 1972, p. 52). Thus, there was no standardization and that made synchronization or integration with others on the factory floor rather difficult or impossible. It also made it difficult for mechanics to obtain spare parts on the market, thus causing undue delays whenever there was a breakdown of equipment that required new parts (see Barker et al., 1991). This was coupled with the fact that the rate of capacity utilization was so low—36.0% in 1978 and in fact, had dropped to 20.0% by 1983 (Central Bureau of Statistics, June 1983, p. 10; SS, June 1989, p. 10).

Also, as pointed out by Chryssides (1974, pp. 267–68) such imported technology used in sawmilling was a major contributory factor to the problem of wastage in the production process, particularly since much of the sawmilling machinery was at least twenty years old. As a result, inadequate sharpening equipment resulted in the production of rough boards of poor quality. Furthermore, a number of firms used circular saws rather than band saws. Since the blades of circular saws are thicker than those of band saws, the former tended to produce much more waste in the form of huge piles and piles of unused sawdust. A further source of such waste generation was from poor sawing through lack of care by workers. When log supplies were poor, the workers took more time and sawed carefully resulting in higher yields, and vice versa. This problem was exacerbated by the shortage of trained saw doctors (i.e., technicians responsible for the maintenance of sawing equipment at the mills) in the country and as a result, saws tended to be frequently badly set and consequently produced inaccurately-sawn boards thus adding to the waste. These patterns—the employment of foreign personnel, the dependence on foreign technology, and the use of imported capital equipment by the industry—discussed above were all tied to patterns regarding the generation of foreign exchange for the country.

Foreign Exchange Generation

As the third most important sector of the Ghanaian economy in terms of exports, the timber sector was not faring well. Specifically, relative to the generation of foreign exchange and in relation to Ghana's balance of payments (BOP), the industry's impact on the country's external debt situation was negative. The importation of capital equipment and machinery together with their spare parts and imports of intermediate products as inputs (particularly, for furniture and fixtures) constituted a heavy drain on the country's very limited stock of foreign exchange, and consequently, a negative effect on its BOP situation. In 1970 and 1975 for example, the wood-sector dominated in the acquisition of import licenses issued and letters of credit established. It accounted for ¢1261.7 million and ¢340.0 million at actual exchange rates respectively. The next and closest sector was petroleum which consumed ¢974.3 million in 1970, and ¢338.0 million in 1975 (World Bank, 1985, p. 26).

According to Combé (1972, p. 116), machinery imported for the industry involved the additional payment of patent fees, with negative implications for the economy. Wood processing, he observed,

> need[s] better and specialized equipment, requiring considerable expenditures. Moreover, they often involve new technologies which are protected by patent

> laws, as is true for example with the mechanized production of mosaic flooring, or certain wood-based panels and of furniture. In all these cases, dependence on foreign exchange is increased because of the need for imported machinery involving additional payment of patent fees. Under these conditions, increased industrialization of the wood-working sector will only provide a limited improvement of the Ghanaian economy, since, at the same time, exterior dependence will also increase (Combé, 1972, p. 116).

Siedman (1971) also pointed out that government revenues were limited by the fact that capital equipment and the intermediate inputs were imported duty-free. In addition, heavy payments in the form of interests on credits and their principals aggravated the BOP situation still further. In 1970, interest payments alone on external public debt was US $12 million and had grown to US $30 million by 1980.

At the beginning of 1983, the interest payment had reached as much as US $38.705 million (Word Bank, 1985, p. 33). Another "leakage" from the country's limited foreign exchange resulted from payments made to the expatriate professional employees engaged in the industry (see figure 3.7). Such personnel were often paid very high salaries with other financial incentives—either wholly or in a high proportion, relative to their Ghanaian colleagues, in foreign exchange. Boateng (1991, p. 30) notes that, to attract such professionals, the companies had to provide very attractive inducements—free, good accommodation, free transport and medical care, plus a good taxable local salary. Any other payment transferred to the foreigner's overseas account (as was often the case) had to be taxed at the rate of 50% however, in line with the income tax law. In a certain case he observed, the effect of the tax on the company whose expatriates' external salaries was as much as 10% of the gross export earnings. This 50% tax raised the proportion to 15%.

Also, prior to the adoption of the structural adjustment program in 1983, freight, insurance, transportation, and travel costs associated with the industry represented the greatest expenditures that influenced earnings potential. These exacerbated the nation's dependence on its foreign exchange, and this was more so with respect to domestic transportation and port conditions, which had been seriously neglected because of economic mismanagement. Generally, Ghana's roads and railways were in a very bad state of disrepair thus leading to frequent breakdown of transport equipment, which had to be fixed with spares purchased abroad, and in other cases replacing such vehicles. The ports were also congested thus causing delays and affecting insurance premiums. Telecommunication facilities had become so poor that it became very expensive to communicate both internally, and with the outside world. All these impacted negatively on the generation of foreign exchange and the balance of payments situation especially, with respect to the temporal costs (delays) and dependence on imports. Combé aptly describes the vicious circle this way: "the importation of foreign equipment, technologies and production methods involve(d) not only the deterioration of foreign exchange assets, but also (made) external dependence so strong that all imported machinery must in some way again contribute to the export trade in order to gain back foreign exchange" (Combé, 1972, p. 114).

In the process of contributing to the export trade, the machinery often broke down, thus creating further demands for machinery and spares and continuing the cycle of dependence thereby.

Figure 3.8 shows the values of inputs—domestic and foreign—from 1977 through 1983 in *Cedis* (¢) the Ghanaian currency. In comparative terms, in 1977, the exchange rate was US $1.00 = ¢1.00. From 1978 to1982, the rate had increased to US $1.00 = ¢2.75, and in 1983, the rate was US $1.00 = ¢30.00. It must also be borne in mind that there is a relatively high proportion of imported inputs in milling and planing as against furniture and fixtures because the former includes the production of veneers, plywood, and chipboards which require a lot of imported chemical inputs. Figure 3.8 also indicates that the foreign component was quite substantial even during those difficult years. This was especially so given the fact that in both sawmilling and the production of furniture and fixtures, the main input is wood, which is obtained domestically. In terms of proportions, the foreign component was as much as 42% in 1977 and before the industry had an infusion of foreign capital after 1983, it declined to 27% in 1981 and to 18% in 1983. This shows that even when foreign exchange was very difficult to obtain, the dependence was still very significant.

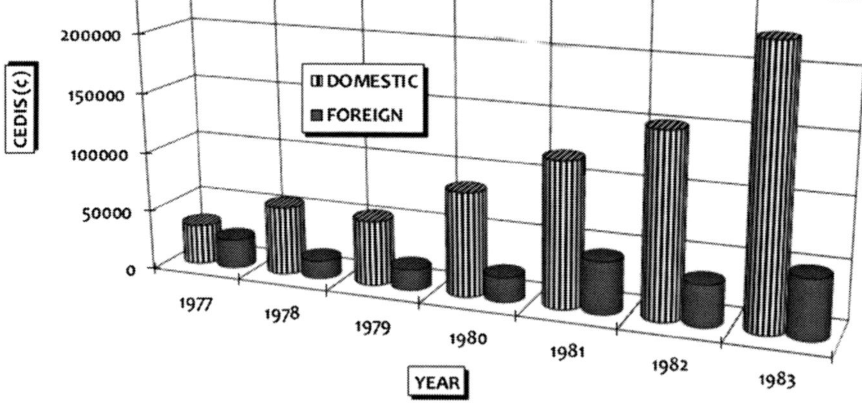

Figure 3.8. Inputs ('000 Cedis): Mill, Furniture, & Fixtures Production (1977–1983)

Table 3.1 shows the grave situation with respect to Ghana's BOP situation, prior to the adoption of the ERP in 1983. The indicators in table 3.1 are in relation to the fact that the average annual percentage of GDP growth was –0.3% for the period 1973–1980, and the per capita average annual GNP growth rate was –2.1% for the same period (World Bank, 1989, pp. 221–22). At the beginning of 1983, the total external debt had even increased to US $1,499.479 million of which US $1,120.638 million and service interest payment for that year was US $38.705 million.

Under such a grave economic situation, this level of dependence on foreign exchange limited or curtailed the remittance of profits abroad because of the need for imported machinery involving the additional payment of patent fees. This was particularly so since besides indirect and other taxes, a tax was collected in order to obtain foreign exchange for each transfer abroad (25% per transfer in the Sixties). Such measures therefore presented a serious obstacle to further investments by foreigners since it ate into the amount of profits they could repatriate (Combé, 1972, pp. 116–17). The taxes on foreign exchange used contributed to the obsolescence of the machinery and equipment in many wood-processing industries because they tended to increase the financial outlay required for such imports. The government's economic strategy then was to keep more foreign exchange in the country's coffers by giving

Table 3.1. Ghana's Balance of Payments and External Debt Characteristics (1970 & 1980)

Year	*1970*	*1980*
Total External Debt	N/A	US $1,312m
Use of IMF Credit	US $46m	US $43m
Short-Term Debt	US $131m	US $108m
Repayment of Principal Debt		
—Public & Publicly Guaranteed	US $14m	US $71m
Total Long-Term Debt		
—Disbursed & Outstanding	US $497m	US $1,138m
—As a % of GNP	23.0%	26.0%
Total Long-Term Debt Service		
—As a % of GNP	1.2%	2.3%
—As a % of Exports of Goods & Services	5.5%	8.3%
Interest Payments on External Public Debt	US $12m	US $30m
BOP & Reserves Current Account		
—Balance Before Official Transfers	US $–76m	US $–54m
—Balance After Official Transfers	US $–68m	US $29m

Source: World Bank (1989, 252–62).
m=million

incentives to indigenous enterprises. This was meant to cut down on the repatriation of profits that would otherwise be the case with foreign-owned firms. The problem with that, however, was that since foreign exchange transactions were regulated and in short supply, the Ghanaian owners also preferred to transfer their profits abroad for the purchase of inputs and for safe-keeping. The patterns relating to firm-ownership, and in fact, the way the industry as a whole was organized greatly influenced foreign exchange generation by the industry.

Firm-Ownership Structure and Organization of the Industry

With respect to the ownership structural pattern, two major developments had taken place since the colonial era. During that period, indigenous Ghanaians were generally excluded from management and ownership of industries either by lack of access to education and/or capital, and even in certain cases, by law (Rodney, 1972). In general therefore, ownership and management were the preserve of Europeans or foreigners from the Middle East and Asia. In fact, the ability of indigenous entrepreneurs to move into larger industries was constrained by the entrenched positions of expatriate businessmen and by the continuing foreign domination of the large companies including the financial ones (Steel et al., 1984, p. 9).

The first post-colonial government of Kwame Nkrumah's Convention People's Party (CPP), as noted in chapter 2, was more interested in the state assuming the dominant role in the industrialization process by directly owning industrial enterprises. This was because Nkrumah did not want to undermine his socialist regime by creating an indigenous capitalist class which would be difficult to control politically (Huq, 1989, p. 139). The development of foreign entrepreneurs was thus encouraged via granting them tax concessions. Nkrumah's preferences in order of importance were as follows:

1. State-owned manufacturing and other enterprises. The State Furniture and Joinery Corporation (currently, Kumasi Furniture and Joinery Company), for example, was founded under this category;
2. Joint State-Foreign and wholly foreign-owned companies, such as Mim Timbers Limited and Saoud Timbers Limited respectively.
3. Intermediate firms which were more labor-intensive as in most furniture-producing firms. These tended to be dominated by Levantine, Indian, and Ghanaian entrepreneurs; and
4. Small-scale enterprises (with less than ten employees) such as carpentry, which were reserved for Ghanaians. Foreigners therefore could not establish firms that employed less than ten people.

On the whole, most of the state-owned operations tended to be over-equipped and operated below capacity. Problems with administration and financial management were compounded by difficulties in production management maintenance and working capital for spare parts. Many machines tended to be cannibalized to provide spare parts for other operating machines. The government was not effective in managing most of these companies and some managers lost or stole millions of dollars through loans and price fixing (Barker et al., 1991, p. 9). The joint state-foreign and wholly foreign-owned companies, on the other hand, tended to be more technically sophisticated, capital-intensive enterprises which also possessed expert management (Huq, 1989, p. 139).

Despite subsequent governments' quest to "capture the commanding heights of the economy" for Ghanaians, an important proportion of the sawmills continued to belong to expatriates. Overall, regarding how both the indigenous Ghanaians and the expatriate owners organized the industry to generate benefits to the country, Combé (1972, pp. 120–21) made the following observations, which had not changed much by 1983:

a. Ghanaians owning small sawmills seem to give absolute priority to their personal gain. They are aware of national problems of forest conservation, foreign exchange, etc., but they rely on government to find solutions.
b. A significant number of expatriate industries are owned by Lebanese, who seem most efficient at their jobs. Although they worked in Ghana's best interest for many years, they nevertheless try to remit their benefits to their own country through complex trade operations with other Lebanese.
c. Public companies whose shareholders are abroad are forced to remit their benefits regularly outside the country. This only augments Ghana's needs for foreign currencies, and the government has fought back by imposing high remittance taxes.

For the pre-ERP period therefore, Combé (1972, p. 121) observed that "it seems that only large national companies are capable of generating significant benefits to the country. Through foreign management and supervision, the expatriate firms have a higher output and better efficiency, but an important share of the profits is remitted abroad, perpetuating the situation of dependence of the Ghanaian economy."

In addition, Siedman (1968, 1971) also pointed out that the foreign-owned firms tended to locate their processing plants in the existing urban centers to take advantage of existing external economies and markets. They also tended to refrain from building integrated plants in competition with their own plants back home. They also did not seek to integrate their projects into the Ghanaian economy which would undermine their own profitable trading activities.

Darkoh (1973, p. 13), also observed that "(t)he inherited commercial and financial institutions also tend to reinforce the pattern . . . The financial institutions (with few exceptions) basically those inherited from the colonial era, . . . have given credit to the large foreign manufacturing firms but seldom to small indigenous entrepreneurs." Thus, the small wood-processing firms did not receive adequate attention and credit facilities from the conventional and often foreign-owned credit institutions (Darkoh, 1973, p. 13).

Furthermore, another major aspect of the problematic structural and organizational patterns characterizing the industry prior to 1983, was the inefficient way the industry's umbrella institution, the Ghana Timber Marketing Board (TIMBOD), was organized and how it operated. According to TIMBOD's Instrument of Incorporation, 1970,[7] it was set up "to regulate the marketing and export of Ghana timber of all kind and species whether log or lumber and to secure the most favorable arrangements for the purchase, grading export and selling of Ghana timber and to assist in the development by all possible means of the timber trade and industry in Ghana" (p. 1). This enabled TIMBOD to control the production and trade of the export-oriented sawmills. The contracts of almost 94% of Ghana's timber exports had to be first approved by the Board and had to correspond to a minimum price level which was calculated in relation with trends on the world market. However, as noted by Combé (1972, pp. 83–83), the minimum price levels were calculated without due regard to the policies of the other African timber-producing countries such as Côte d'Ivoire, Cameroon, and Nigeria, where such controls were absent. For quite a number of times, Ghana was in a totally isolated position, thus disturbing its situation on the world market and stymied the expansion of its timber production. By restricting the larger producers from making foreign sales at lower prices, TIMBOD often caused the depreciation of processed material in stock, thus decreasing foreign exchange earnings. This was in contrast with the rapid growth in the neighboring Côte d'Ivoire, for example, which had no such regulations.

As shown by Chryssides (1974, pp. 77–80), the vetting of contracts involved unnecessary delays for interested parties. The Board only needed to check contract prices against the relevant minima on its price lists. Contracts however took some three to four weeks to be approved. Consequently, prospective buyers who urgently needed lumber turned elsewhere for their supplies. Ghana was thus not considered a worthwhile source of lumber. Contracts were also lost when the Board's minimum price level proved unrealistic. The minimum price for each grade and species of wood was determined by the average price of lumber accepted during the previous quarter. The Board therefore could not respond promptly to changes on the world market. When world prices were falling, prices acceptable to the Board tended to be totally uncompetitive. Conversely, when prices were rising, the exporters shipped their wood at artificially low prices, and kept their profits overseas—a situation which the minimum price system was intended to prevent. The same results were achieved when large contracts were reached when prices were lowest, particularly by large international companies, which sent regular large supplies to associated plants overseas. Also, owing to a lack of an organized market intelligence system to provide weekly information on current prices of lumber on the main buying centers of the world, old contract prices were used to determine the acceptability of new contracts. This made the whole process inefficient.

Finally, regulations relating to the minimum price system could be easily circumvented. There were over forty-six grades of timber species, and the Board's price list for lumber included separate figures for each grade, and no attempts were made to check whether the quality of lumber corresponded to the specifications of the various contracts. The regulations required that the surface of individual pieces of lumber be examined—a difficult process, since that involved undoing bundles. Consignments were thus not inspected. Parties to contracts were therefore free to make declarations regarding the quality of the lumber and so were able to export below official minimum prices for higher grades shipped. This was particularly more so considering the fact that current contract prices were used in establishing minima for the next quarter. Falsified contract prices therefore caused the Board's minima to depart progressively from reality since each more unrealistic price figure in turn gave rise to further falsification. According to Combé (1972, p. 83), the most pressing concerns of the firms were the high purchase and maintenance costs of equipment, but TIMBOD gave no attention to them in that regard. In essence therefore, the Board was not "rendering any useful service to the industry, and indeed had been no more than a levy-collecting agency and dispenser of funds" (Chryssides, 1974, p. 74). To many sawmillers and exporters, the Board was characterized by corrupt practices whereby, by means of bribes, exporters could have their contracts approved though the prices were below the official minima (Chryssides, 1974, p. 80).

The above pre-1983 problematic structural and organizational patterns are attributed, on a broader scale, to the political/economic context—national, regional, and global—within which African economies operate, according to Sawyerr (1990, pp. 9–10). The context, as he points out, was "defined by colonial experience and the international relationships that were established thereby; as well as the relationship of both to contemporary domestic and external economic structures." In order to examine the relative changes emerging in the structural and organizational patterns of the wood-processing sector relating to the ERP, there is the need for us to discuss the nature of neo-liberalism and scope of SAPs in SSA, and situate the sector in that context, and this is done in chapter 4.

NOTES

1. See also, *Plunder in Ghana's Rain Forests for Illegal Profit.* Special Report, Vols. 1 and 2. London, Friends of the Earth.
2. *Government Policy on the Promotion of Ghanaian Business Enterprises*, 1968, Ministry of Information, Accra, P. 1.
3. Mr. Desmore, the original owner of Mim Timbers Ltd., was naturalized as a Ghanaian citizen and continued to manage it as he did before. Mim Timbers Ltd. therefore did not experience the problems that characterized the other government-owned companies.
4. The Statistical Service of Ghana lists items under ISIC codes 3311 and 3320, as Sawmills, Planing and Other Woodmills, and Manufacture of Furniture and Fixtures except primarily of metal respectively. Combé (1972, 46) notes for instance, regarding the period preceding the mid-seventies, that "unfortunately, statistical data was not available in a sufficiently detailed form, since agriculture, forestry and fishing are generally combined. Similarly, sawmills and plymills are considered as a single manufacturing group in statistical handbooks."
5. This is aside from the fact that the quality of a log may determine the proportion that has to be cut out based on the defects and some logs have more defects than others. The diameters of the various species are also a determining factor in that sawing small logs involves proportionately greater loss of volume in the log squaring process.
6. These were locations outside the main and limited grid of the Akosombo Hydroelectric Power Plant which was built mainly to generate power for the local aluminum smelter at Tema but had transmission lines reaching Kumasi and Takoradi at their then extremities.
7. *Ghana Timber Marketing Board Instrument, 1970*, Legislative Instrument No. 661, Accra: Ghana Publishing Corporation.

Chapter Four

Turf Reclamation Strategies

Neo-Liberalism and the Timber Industry

Beginning in the early 1980s, most countries in Sub-Saharan Africa (SSA) including Ghana succumbed to Neo-Liberalism, an economic paradigm, which metamorphosed from economic liberalism in the 1970s onwards among some economic scholars and policy makers. According to the renowned geographer and social theorist, David Harvey (2005, p. 2),

> Neoliberalism is in the first instance a theory of political economic practices that proposes that human well-being can best be advanced by liberating individual entrepreneurial freedoms and skills within an institutional framework characterized by strong private property rights, free markets and free trade. The role of the state is to create and preserve an institutional framework appropriate to such practices. The state has to guarantee, for example, the quality and integrity of money. It must also set up those military, defense, police and legal structures and functions required to secure private property rights and to guarantee, by force if need be, the proper functioning of markets. Furthermore, if markets do not exist . . . then they must be created, by state action if necessary. But beyond these tasks the state should not venture. State interventions in markets (once created) must be kept to a bare minimum because, according to the theory, the state cannot possibly possess enough information to second-guess market signals (prices) and because powerful interest groups will inevitably distort and bias state interventions (particularly in democracies) for their own benefit.

Dag Einar Thorsen and Amund Lie (n.d.) in the same vein, note that generally, neo-liberalism includes the belief that freely adopted market mechanisms is the optimal way of organizing all exchanges of goods and services as well (see Friedman 1962, 1980). Free markets and free trade will, they believe, set free the creative potential and the entrepreneurial spirit which is built into the spontaneous order of any human society, and thereby lead to more individual liberty and well-being, and a more efficient allocation of resources (see Hayek 1973; Rothbard [1962/1970] 2004).

As an ideology championed by the two foremost international financial institutions—the International Monetary Fund (IMF) and the World Bank—in SSA and the Third World in general in connection with these developing countries' debt crisis, neo-liberalism in essence promotes and facilitates free trade between nations and privatization and the retreat of government activism within nations. It does this by ensuring unfettered movement of goods, resources, and investment capital in the quest by international capital to access cheaper resources, and exploit such resources with the maximum efficiency in order to maximize profits. It is in relation to this that Gwynne et al. (1999) noted, as was pointed out in chapter 1, that the impact of developing countries being more fully inserted into the global economy, increasingly reduces their room for policy maneuver, as their governments become more dependent on the policy approval of global institutions—such as the IMF, the World Bank, and the WTO—that "supervise" the world economy and on the investment decisions of multinational companies that can be swayed by the verdicts of those international institutions. Neo-liberalism thus constitutes an externally imposed paradigm reified as a strategy to address the structural issues in the developing economies of SSA, given as we noted earlier in the preceding chapters, that it is the structure of African economies that generally defines the essential features of the region's central problem of underdevelopment (World Bank, 1981; OAU, 1982; UN/ECA, 1986).

Briefly returning to our discussions in the previous chapters as a link to neo-liberalism, we noted *inter alia*, that:

1. For most scholars, analysts, and practitioners of development in the Third World, structural problems generally refer to the social and economic features of an economy, which impede development (Engberg-Pedersen et al., 1996, p. 3); and
2. In SSA economies, the major structural constraints include such features as the externally oriented nature of the region's economies, particularly the production and export of primary products; and the economic (over)dependence of the region's economies on the advanced industrialized economies, all of which are legacies of the colonial era (see UN/ECA, 1986).

It was therefore to address the structural problems dogging these economies, at least in theory, that almost all the ailing economies of SSA succumbed to neo-liberalism by adopting and implementing some form of a structural adjustment program (SAP) at the behest of the two powerful Bretton-Woods institutions namely the IMF and the World Bank for an "accelerated" and "sustainable" development in that region (see World Bank, 1981, 1989). The acronym, SAP, and the term, adjustment, are used interchangeably in our discussion. They refer to the economistic, neo-liberal, or orthodox structural adjustment programs promoted by the Bretton-Woods financial institutions vis à vis the relatively holistic "African Alternative Framework to Structural Adjustment Programs for Socio-Economic Recovery and Transformation" (AAF-SAP), which was advanced by the United Nations Economic Commission for Africa (see UN/ECA, 1986).

In contrast with the view that Africa's structural problems are largely legacies of the colonial era, for the international financial institutions (IFIs) like the IMF, the World Bank, and the neo-liberal proponents of adjustment, however, these constraints on development are largely the result of a single structural problem, namely market distortion (see World Bank, 1981; Green 1987, 1991; Steel, 1988; Younger, 1989). Market distortion, in this case, is attributed to undue state intervention in the region's economies. Accordingly, "the root cause of Africa's current ills lies not in its history of colonialism and its form of integration into the world economy, but in the policies which have aimed to offset these imbalances, with their unintended effects in the form of excessive state intervention in the economy" (Engberg-Pedersen et al., 1996, p. 4). Consequently, the neo-liberal structural adjustment was "needed in order to restore the long-run conditions for stable economic growth," and that "an intermediate period of austerity should be self-imposed, so that market mechanisms may at last come to equilibrium" (Schwarz, 1990, p. 23). This elicits a discussion of the nature of adjustment and its relevance to structural transformation in Africa.

THE NATURE OF STRUCTURAL ADJUSTMENT PROGRAMS AND AFRICAN DEVELOPMENT

SAPs are a set of economic policies and prescriptions designed, promoted, and sponsored by the IMF and the World Bank to address the perceived market distortions by "getting the prices right" in Third World regions like SSA, as a strategy for forging development (World Bank, 1981, 1989; Steel, 1988; Younger, 1989; Green, 1987, 1991). In general, SAPs, as per Harvey's (2005) definition of neo-liberalism, entail rolling back the interventionist role of the state in favor of a free and competitive market system. The government is relegated to the role of creating an enabling environment for investments by local and foreign capital. Generally, policy measures under SAPs involve all or some combination of the following (see Sawyerr, 1990, pp. 20–23):

1. reducing public expenditure to reduce public borrowing and budget deficits and shifting the pattern of public expenditure to the export sector; increasing domestic savings to promote investments;

2. reorganizing the state-owned sector to make it competitive, self-supporting, and profitable;
3. liberalizing the economy to allow the free forces of the market to determine prices;
4. promoting exports to generate foreign exchange through an incentive system; and
5. promoting foreign private investment through the promulgation of investment codes with extensive concessions to foreign investors;
6. the provision of foreign loans to rehabilitate malfunctioning economic sectors; and
7. the rehabilitation of physical infrastructure to smooth the production and transportation processes.

This package normally includes massive devaluation of the local currency to improve the competitiveness of exports, with the object of generating a trade surplus to meet the adjusting nation's debt servicing obligations. In sum, SAPs entail a combination of three main measures, namely the stabilization, rehabilitation, and liberalization of the adjusting economy, resulting in the retreat of government activism in favor of a free and competitive market system and the creation of an enabling environment for investments by local and foreign capital. As a neo-liberal development strategy, adjustment "is based on the postulate that economic efficiency and rational management along the lines of market mechanisms should constitute the sole scale upon which all national considerations of production, administration and culture are measured" (Ould-Mey, 1996, p. 16). Under neo-liberalism, as noted by Gwynne and Kay (1999, pp. 12–13), the adopting state must be willing to sacrifice uncompetitive sectors—most notably in industry—to foreign competition, a corollary of which is a return to relying on natural resource advantages—tropical timber in our case—and what has come to be known now as non-traditional exports like sliced veneer and poles. To the sponsors, therefore, SAPs constituted the best trajectory for addressing Africa's development problems (Green, 1987, 1991; Steel, 1988; Younger, 1989). Unlike most of the adjusting SSA countries, given the state of the economy and the political conditions of the time, Ghana embraced and implemented the program as prescribed by the sponsors in April 1983. What then was the state of the economy? What prevailing political conditions at the time made adjustment attractive and of what relevance was the timber industry to the decision to succumb to neo-liberalism?

GHANA'S PRE-SAP ECONOMIC AND POLITICAL CONDITIONS, AND THE TIMBER SECTOR

Ghana's economy was one of the most promising, in terms of development, in SSA at the time of independence. It was arguably the most economically advanced of the British colonies, a "middle-income country and the world's leading producer of cocoa" (Engberg-Pedersen et al., 1996). Its constituent timber sector had historically accounted for 5–6% of Ghana's total GDP and 11% of total commodity export earnings. Wood from this sector was the third most important item after cocoa and minerals among commodity exports. In addition, forestry was and is a renewable resource-based sector with immense potential for upstream and downstream linkages and employment. In the 1970s, however, Ghana's economy along with its constituent timber sector took a downturn and reached its lowest ebb in the early 1980s with serious political repercussions (see Ninsin, 1989).

In the specific case of the forestry sector, it had declined to its lowest ebb by 1982 (just before the ERP). In 1977 the sector employed 20,900 people but by 1982 employment had declined to 14,442 people. Also in 1977, 1.3 million m^3 of wood from the sector were exported, fetching US $136 million. By 1982, the volume and income figures had declined to 100,303 m^3 and US $15.3 million respectively. A number of factors led to its decline. By 1983 the basic support infrastructure had largely been neglected. For a long time the roads and railways had been poorly maintained and most sections were in very poor condition. The ports had also not been upgraded and had become very inefficient in terms of operations and facilities available (e.g., no container-handling facilities). The telecommunications system had also been neglected and was extremely unreliable. This neglect greatly increased the cost of business operations by the wood processing firms (World Bank 1989) which translated into extra costs for firms (depending on their geographical location). It must be noted that the tropical hardwood species which were and continued to be exploited primarily for export have long gestation periods. For example, an average-size *Odum* tree (*Chlorophora excelsa*) takes about two hundred years to attain harvestable size. Others like *Afromosia* or African Teak (*Pericopsis elata*) take as long as five hundred years to attain a radius of three feet (see Owusu, 1994). Yet, despite the heavy export volumes involving heavy collateral environmental damage, tree replanting was and continued to be insignificant. Also, processing the wood for export involved a lot of waste. This is aside from the fact that the quality of a log may determine the proportion that has to be cut out based on the defects, and some logs have more defects than others. The diameters of the various species are also a determining factor in that sawing small logs involves proportionately greater loss of volume in the log squaring process. Wood recovery rates were very low, averaging 42% (Chryssides 1974), hence for every 100 m^3 of processed wood exported, 58 m^3 more was wasted.

All these were largely the result of prior economic mismanagement (Huq, 1989; Rado, 1985). Table 4.1 shows some of the distortions emerging in the general economy from 1970 through 1980 after some growth in the 1960s. Between 1979 and 1982, Ghana experienced a negative annual average GDP

Table 4.1. Some Indicators of Growth, Efficiency, and Distortions (1960–1980)

Indicator & Period	*Average Annual Growth Rates*
1. GDP 1960–1970	2.1%
2. GDP 1970–1980	0.2%
3. Agriculture 1970–1980	–1.2%
4. Industry 1970–1980	–1.2%
5. Per Capita Income 1970–1980	–3.1%
6. Export Volume 1970–1980	–8.4%
7. Average Rate of Inflation 1970–1980	34.8%
8. Gross Domestic Investment 1970–1980	–5.0%
9. Real Rate of Interest 1970–1980	–30.0%

Source: World Bank (1984, 3).

growth rate of -6.1% (World Bank, 1984, p. 3). By the beginning of 1983, the economy was heavily in debt and in fact, had sunk into a very serious economic malaise, which was described in the September 3, 1989 issue of *The Economist* (p. 40), as "an extreme case even in Africa."

Table 4.2 shows Ghana's strangling external debt and service payments from 1980 through 1983. At the beginning of the 1983 adjustment year, the outstanding external debt was US $1,120,638,000 (disbursed) and US $1,499,479,000 (undisbursed) with a total interest payment of US $38,705,000, and meeting the debt servicing obligations was thus in jeopardy. As noted by Engberg-Pedersen et al. (1996, p. 11), the "two decades of financial laxity and mismanagement had led to macroeconomic chaos. Exchange-rate overvaluation and high export-taxation led to rapidly declining exports (cocoa, timber, gold) from the mid-1970s, while external factors plus a large public spending deficit increased the rate of inflation to an average 80% per annum." Consequently in the early 1980s the Ghanaian government, like many others of the heavily indebted developing world, began to lose the option of doing business with foreign lenders

Table 4.2. Ghana: Pre-Adjustment External Public Debt and Service Payments (1980–1983) in Thousands of US Dollars

	Outstanding Debt at Beginning of Period		*Service*		
Year	*Disbursed Only*	*Including Undisbursed*	*Principal*	*Interest*	*Total*
1980	983,899	1,403,638	65,553	28,382	93,935
1981	1,100,426	1,534,974	28,255	24,966	53,221
1982	1,114,020	1,503,523	34,399	27,778	62,177
1983	1,120,638	1,499,479	61,041	38,705	99,746

Source: World Bank (1985, 33).

on an exclusively bilateral basis. To ensure control of their turfs, the promotion of SAPs therefore became a concerted effort on the part of lender countries and institutions to deal with borrower countries on a case by case basis rather than as a group. The programs were implemented with a strategy of providing loans to the indebted developing countries and dictating policies. Ghana, like other indebted developing countries, was simultaneously seduced by the foreign loans, frightened by its fiscal crises and balance of payments deficit, and haunted by the possibility of being declared insolvent by the IMF and the World Bank (Ould-Mey, 1996).

In relation to the economic crisis, the prospect of insolvency, and the government's political survival, Ninsin (1990) shows, as we shall soon see in the political economy of Ghana's structural adjustment, that the Ghanaian government's adoption of the ERP was essentially metonymical rather than a genuine measure for fashioning and sustaining development. It was metonymical on the part of the government in the sense that it was used in tandem with their development objectives to promote and achieve its hidden interests and agenda (see Gore, 1984, p. 258). In the main, the nature of the macro-economic policy measures and the specific policies and actions of the program, as in the case of the formal wood processing industry, rather feathered the nest of the Provisional National Defence Council (PNDC) government and simultaneously met some demands of international capital. According to Ninsin (1989), the program was adopted as a convenient measure to simultaneously address a political and economic (spatial organizational) problem. More specifically, it was a convenient measure to defuse a domestic political and economic crisis the government itself helped to fashion and to stave off its own imminent collapse as a consequence of its initial failure to "accommodate international capital," by its socialist anti-West and populist stance. To shed more light on this, there is the need to look at the competing perspectives on adjustment, and the political economy of Ghana's adjustment program, dubbed the Economic Recovery Program (ERP).

COMPETING PERSPECTIVES ON NEO-LIBERAL STRUCTURAL ADJUSTMENT PROGRAMS

Two main but competing perspectives are identifiable with respect to the developmental role of SAPs in SSA. The first is that of the Proponents, which represents mainly the neo-liberal views of the IMF/World Bank Internalists; and second, that of the Critics, composed of World Bank in-house criticism on the one hand, and radical or structuralist criticism, which is more concerned with the nature of the process whereby economic growth and development are achieved; and indigenous control of national destiny.

For the Proponents, SAPs, the "invisible hand of the market," constitutes the surest path to removing the bottlenecks in the economy, because it makes production, exchange, and distribution in economies more efficient through competition (see Storper and Walker, 1992, p. 36). They proceed from the basic neo-classical/liberal premise that Africa's development problems were primarily internally generated and sustained (World Bank, 1981; Green, 1987 and 1991; Steel, 1988; Younger, 1989; and Schwarz, 1990). The solution then has to begin internally, employing the principle of comparative advantage in the production and export of natural resources and must be operationalized with international capital, since "growth with debt" (such as the use of sector adjustment loans or SECALs) enables the borrowers to increase their capacity and to expand with the aid of net foreign savings. They argue that SAP policies involving the improvement of the incentive system, rehabilitation of the basic infrastructure, and the creation of a favorable economic environment to expand production and exports *inter alia*, according to the proponents, should:

1. enhance production linkages by strengthening and increasing the connections between the primary-stage processing firms and the suppliers of non-primary inputs including equipment and spares. This should make transportation and communications much easier; and
2. promote service and marketing linkages to the primary-stage processing firms, by making it easier for the providers of such services as financial, maintenance, and technical inputs, to link up better with them.

As far as the operations of firms are concerned, they believe that program weeds out inefficient firms or those which integrate in an inefficient way since the market forces will force them to fold up or devise new strategies to deal with the competitive, volatile, and uncertain environments (Dicken and Lloyd, 1990, p. 305). It is also expected by the proponents that devaluation of the "overvalued" local currencies will result in the reduction of imports, as to favor a reallocation of factors of production to the sectors producing tradable goods and services (Schwarz, 1990, p. 23). Industry, the supporters argue, suffers because regulation and controls inhibit the free flow of capital through the economy (Hamilton, 1989). These expectations are, however, disputed by the critics, particularly the radical political economists.

Unlike the Proponents, the Critics' position begins with the premise that the problematic structures are the consequence of the incorporation of the region's pre-capitalist economies into the global market system (Slater, 1975; Mabogunje, 1981; Sawyerr, 1990). For example, in the case of Ghana, the structural problems are thought to have emerged from the way Britain organized the Gold Coast as a tributary colony to expand her own economy, leaving a legacy of a "warped" economy with "untransformed structures," an economy with a one-sided external orientation and dependence (see Sawyerr, 1990). Accordingly, SAPs are seen as a mechanism for regaining their turfs by opening up African economies and their markets to Western business interests a little more (Hamilton, 1989, p. 1529). In this regard, the technological requirements for the rehabilitation of primary-stage processing industries like that of timber, compels adjusting countries to purchase production, transportation, and telecommunications equipment, as well as their spare parts, thereby reinforcing the dependence on imported technology and to some degree, on foreign technicians and other related personnel. The procurement of such equipment and spares is normally tied to bilateral and multilateral loans schemes such as the SECALs, a process aided by the reduction or removal of foreign exchange controls, the lifting of restrictions on imports, and the free repatriation of profits by foreign investors (Sawyerr, 1990).

To the critics therefore, SECALs worsen balance of payments situations, and force firms to produce more, at lower prices, to defray debts and service charges. The procurement and debt repayment processes are, therefore, seen as mechanisms for supporting firms and the economies of the donor countries. Hancock (1989, pp. 156–70) noted that Britain, for instance, normally allocates some £850 million per annum to its bilateral aid programs. Out of that, some 80% is typically used by the recipient countries to buy British goods and services. He documents that according to the Overseas Development Administration (ODA), multilateral agencies can normally be relied upon to purchase British goods and services with a value equivalent to 120% of Britain's multilateral contributions. For the U.S., he noted that, through the World Bank, 82 cents of every tax dollar that is contributed by the American tax-payer are immediately returned to American businesses in the form of purchase orders. The Critics also argue that the overdependence on expensive imported technology and equipment makes matters worse by the retrenchment of workers in industries like that of timber, in the name of efficiency, despite the huge reserves of unutilized and underutilized labor. In respect of that, Barker et al. (1986, p. 7) suggested that productive resources available to such industries are not fully utilized as required in the Ricardian sense of comparative advantage. The flexible exchange rate system, emphasized under the program, *inter alia*, hurts industries (Sawyerr, 1990, p. 31) because:

i. It has the tendency to push the price of foreign exchange beyond the reach of local entrepreneurs who want to import capital goods or equipment, thereby biasing the pattern of imports away from such essential imports; and
ii. It floods the market with finished goods and other inputs that could otherwise be produced domestically.

Such imports result in a decline in the demand for the import-competing output of firms producing related raw materials, reduce linkages, and kill potential ones. In addition, it reduces industrial capacity and deepens the external dependency of industry. This is due to the fact that such imports tend to have a ready market and thus yield high profits more quickly. Importers of such commodities are thus able to generate huge amounts of the (local) currency and therefore outbid the importers of the essential machinery and industrial equipment. Their competition also tends to bid up the price of foreign exchange (Sawyerr, 1990, pp. 32–33).

Also regarding the infrastructure rehabilitation measures under SAPs, the critics note that the restoration or development of basic infrastructure has a heavy bias towards infrastructure related to export production. Sawyerr (1990, p. 30) for example, argues:

> harbours, railways, trunk roads, etc., have tended to attract a disproportionate amount of the new public investment. A strong argument could be made that the restoration of such infrastructure is indispensable to economic recovery at any event. It just happens that this rebuilding of infrastructure tends to reinforce the commitment to, and dependence on, the export of primary products. While the assured supply of such products certainly benefits the industrialized economies, the benefits to the SSA countries of continued and increased production of primary commodities for export, are less clear in the long run, but even in the short run, the already low export prices should give cause for worry. Moreover the nature of investments in such infrastructure and the procurement methods adopted in connection therewith invariably involve the importation of machinery and other finished goods, contractors and consultants from the Western industrialized countries.

Such measures under SAPs favor the Western industrialized countries more than they do the SSA economies by reinforcing the former's control over, and even expand, their turfs in the latter. In fact, in the case of the SSA economies like Ghana, even in the export of primary-stage processed natural resources, which are supposed to gain more value in Western markets, the unfavorable production and market conditions are almost the same as primary products. Roemer (1981, p. 432), for example, argued that refining industrial raw material for export (as in the production of lumber and furniture parts based on the buyers' specifications) may increase the risk of adverse external shocks. This is because the exports then become "specific products" produced by "specific techniques" and thus become subject to greater risks of substitution by new products or obsolescence in the face of new technologies.

As regards external linkages, the focus of geographic linkages has traditionally been abstracted from its broader context, namely, the asymmetrical power relationships that exist between firms located in the peripheral regions as SSA and those in the Center, much as geographic space is separated out to exclude other dimensions by neo-classical economics. Linkages, therefore, have tended to be perceived in purely spatial terms with emphasis on geographic proximity (see Dicken and Lloyd, 1990, pp. 332–33). The negative effects of the asymmetrical power relationships pertaining to processing raw materials for the export market thus cannot be ignored under SAPs. This is because sawmills and the firms which produce lumber and furniture-parts respectively for export, for example, under the program constitute "process and components subcontractors" relative to the "principal firms" overseas. In other words, they engage in "speciality" and "cost-saving" subcontracting. In this regard, the asymmetrical relationship between the principal firm overseas and the subcontractors may be short- or long-term, or in certain cases, single batch. The principal firm may also provide detailed designs or specifications, or in other cases provide technical, finance, general assistance, or advice. Under this "just-in-case" system, Sayer (1986, pp. 47–48) notes:

> Both functionally and geographically, the relationship between firms and their suppliers tends to be distant; how suppliers operate is of no concern to the purchaser, provided the price is right, and this lack of inter-firm contact together with the infrequent nature of deliveries, allows suppliers to locate . . . considerable distances away from the purchaser, if by doing so they can minimize labor costs and other expenses.

Such firms therefore stand at a disadvantage in cases where the principal firms engage in dual or multiple sourcing and play them against one another with respect to prices. Also, because export processing demands efficient production, it does not have the protective feature of import substitution that may help develop neophyte entrepreneurs and managers (Roemer, 1981, p. 431).

Mytelka (1989, p. 107) aptly sums up the critics' position regarding SAPs as a development strategy in SSA. She noted, "[S]tructural adjustment programs that lead to more debt are thus not a solution. They are part of the problem. It is time therefore to move beyond mere palliatives, such as increased bank lending or direct foreign investment flows, to a more fundamental rethinking of the mass production model." To the Critics, therefore, SAPs simply just do not address some basic requisite structural issues. They point out that the state is indispensable in addressing those basic structural problems; and the panaceas for addressing the structural problems should take cognizance of the unique history and the specific context of the adopting economy (UN/ECA, 1986; Mytelka, 1989). It is in relation to the unique history and the specific context of Ghana's economy that we now turn to the political economy of Ghana's SAP. To do that, let us look at Ghana's Economic Recovery Program (ERP) as the country's version of the IMF-World Bank sponsored SAP.

THE ERP AS GHANA'S SAP

The ERP as Ghana's SAP constitutes the general economic framework within which the PNDC government sought to address the problematic patterns by creating appropriate structures and an enabling environment *inter alia*, for more efficient production and distribution in the ailing Ghanaian economy. As a comprehensive measure for addressing Ghana's economic crisis and for that matter, the timber industry, the ERP was initiated in April, 1983. It was not simply a package of ameliorative economic policies implemented for that year alone, but rather, a continuing approach to stabilizing, rehabilitating, and restructuring the Ghanaian economy through a process of economic liberalization. As pointed out by the then Head of State and Chairman of the PNDC, Flt. Lt. J. J. Rawlings, "Our basic objective is to establish the right environment to do business and to make investments; both foreigners and our own citizens can then take advantage of the environment" (GIC, 1990, p. 7).

The ERP was thus a serious attempt at addressing issues relating to the proper management of macro-economic and structural transformation policies for a productive and efficient economy to satisfy the needs of both Ghanaians and international capital. This is because, in the final analysis, it is political stability and the prevalence of more conducive macro-economic conditions that determine private sector participation in an economy. Three broad phases with respect to the ERP are identifiable: The first phase focused primarily on "Stabilization" and covered the period 1983–1986. It emphasized "getting the prices right" in the Ghanaian economy, as well as reducing the severe imbalances in government finances and restraining credit expansion. In connection with those measures, the government undertook large discrete devaluations of the exchange rate, lifted domestic price controls, mobilized its revenue by broadening the tax base, strengthened tax administration, and tightened control over its expenditure.

In addition, there was an emphasis on increasing the productive capacity of the export sector by rehabilitating industries that were capable of quickly generating much revenue for the government using an attractive incentive system as in the case of formal wood-processing. Also, removal of infrastructural constraints particularly in the transport sector received priority attention. At the Donors' Conference attended by such important industrialized countries like Canada, France, Germany, Italy, United Kingdom, USA, and financial institutions like the European Investment Bank and the World Bank, in November, 1983, for instance, it was decided that two-thirds of the government's foreign exchange expenditure over 1984–1985 was to be spent on rehabilitating the railroad network, roads, and the Tema and Takoradi ports. In 1984, $860,000 was provided by the government for the rehabilitation and development of posts and telecommunications out of $13 million under the program (Government of Ghana [GOG], 1984, p. 13).

The second phase was primarily one of "Rehabilitation," and covered the period 1987–1989. In this phase there was a substantial increase in government investment outlays for rehabilitation of the economic and social infrastructure. In addition, the government initiated structural changes to address the structural roots of the economic imbalances and to rebuild the productive base. The government thus focused on comprehensive liberation of the exchange and trade system. It embarked on divesting itself from the state enterprises sector, reforming the civil service by reducing its size, and increasing remuneration levels in relation to inflation. It continued with the tax reform process and tax administration improvement, and engaged in institutional and financial reforms to strengthen the banking system.

The third phase covering the post-1990 period, focused mainly on the "Liberalization" of the economy. The period was marked by an acceleration of structural and institutional reforms to achieve a more even balance between the macro-economic adjustment and structural policies. Decisive steps were taken towards a market-oriented economy by the completion of the exchange rate reform, increased reliance on market-based instruments of monetary policy as well as significant improvements in the fiscal incentives for savings and investment by the private sector. Finally, the period saw the reinforcement of monetary management and intensification of efforts to reduce the level of inflation to a low and stable level. To put these changes in context and make them meaningful, let us now turn to the political economy of Ghana's SAP.

THE POLITICAL ECONOMY OF GHANA'S STRUCTURAL ADJUSTMENT PROGRAM

As we noted in the previous chapters, Ghana's economy is heavily dependent on exports to the developed world; hence we need to situate our discussion in the context of the broader global economy. Our discussion is, therefore, predicated on the structuralist view that the Ghanaian political economy, and for that matter, its structural and spatial characteristics—in terms of production and exchange—resulted primarily from its incorporation into the global capitalist system (Frank, 1969; Wallerstein, 1974; Slater, 1975). The process of incorporation, we noted, entailed a formal political and economic organization of Ghana's pre-capitalist economy into a primary export one. Consequently, Ghana's exports are now oriented to the demands of the dynamic "Centers" of the global capitalist economy primarily, Europe and to a lesser extent the U.S. and Japan. The incorporation process, described by Frank (1969) as "the development of underdevelopment," had two dimensions: on the one hand, the external integration into the global economy; and an internal disintegration (Slater, 1975) as reflected in some of the patterns in the export-processing industry such as dependence on external capital, technology, and markets, and the lack of production linkages, or downstream processing (secondary- and tertiary-stage production) internally, on the other.

We have also noted in chapter 1 that as a result of the incorporation process, post-colonial governments of Africa have, as a legacy of the colonial past, had to choose between two options with respect to relations with the Center (Ninsin, 1989). The first is to sustain the asymmetrical relationship in the global capitalist system as a peripheral economy, together with its superstructural political apparatus dominated by the national bourgeoisie. This option, we noted, demands a mandatory accommodation of international capital. The second is to delink or rupture the economic and other links with international capital. According to Samir Amin, one of the main proponents of delinking, the measure does not mean autarchy. Rather, it is "the refusal to submit to the demands of the worldwide law of value, or the supposed 'rationality' of the system of world prices that embody the demands of reproduction of worldwide capital. It presupposes the society's capacity to define for itself an alternative range of criteria of rationality of internal economic options, in short a 'law of value of national application'" (Amin, 1990, p. 71). The purpose of this option then is to create a new structural basis and impetus for autonomous national development. This approach involves a radical restructuring of the old socio-political relations. However, as noted by Tandon (1985), with this option, there is a real prospect of collapse if it lacks a firm socialist vis-à-vis a mere popular content, given the resilience of the capitalist law of value.

These two options confronted post-colonial governments of Ghana in the face of the declining economy, and with implications for the timber industry, and were thus instrumental in shaping the policies and actions designed to arrest the development problems prior to 1983. They also informed and influenced the Provisional National Defence Council (PNDC) government's policy measures under Ghana's SAP, the ERP. It is thus within that context that the ERP is seen to be metonymical.

The ERP as a Metonymical Measure

The rise of J. J. Rawlings to power, both as the leader of the Armed Forces Revolutionary Council on June 4, 1979 and of the PNDC on December 31, 1981, is explicable only within the context of Ghana's political and economic unrest. Rawlings and the PNDC government, unlike the previous administrations, "correctly comprehended the dual character of the crisis: political and economic" acknowledging that "the first arose from the second" (Ninsin, 1989). Rawlings thus realized, from the experiences of the previous regimes, that the justification and legitimization of the PNDC regime politically, rested practically on the restoration of the country's economic health, after running into problems, by tackling the political character first with his initial anti-West socialist and populist stance. Given the resilience of the capitalist law of value, his regime thus faced the real prospect of collapse because his stance had a mere popular content and lacked a firm socialist content as noted by Tandon (1985). To reverse the order, he pragmatically had to accommodate international capital. Hence, the colonial "old wine" of producing raw materials and semi-processed items (such as logs and lumber) for export was, under the circumstance, indispensable. This "old wine" however, had to be put in a "new wineskin," namely succumbing to neo-liberalism through the adoption of an SAP under the auspices of international capital through the IMF and the World Bank.

Unlike the Rawlings regime, the People's National Party (PNP) administration of Dr. Hilla Limman, which was constitutionally elected in August 1979, did nothing, in its time, to address the then explosive economic and political situation. Essentially, it pursued the same policies as General Acheampong's National Redemption Council government by refusing to accommodate international capital. This it did by refusing to devalue the *Cedi*, reduce its borrowing sharply, or until shortly before its overthrow, to raise significantly the producer price for cocoa, thus deepening the social and economic crisis (Huq, 1989, pp. 16–17). Consequently,

> [w]orkers and progressive organizations responded to the damaging social effect of the economic crisis with sustained and widespread political agitation and open denunciation of the government. It was clear that parliamentary (liberal) democracy had failed to open the domains of power to the popular masses as a way of giving public policy a popular content. Its only justification, it seems, was that it was appropriate for promoting bourgeois rule and alienating the masses from the instruments of rule as well as from the material means of labor and of life (Ninsin, 1989, p. 135).

It was, therefore, under those circumstances that Rawlings ousted the PNP regime, and ignorant of the consequences of refusing to accommodate international capital, proceeded immediately to first resolve the political crisis by disorganizing the bourgeois institutions of rule and then reconstituting them to incorporate the masses. This was an attempt to delink with a mere popular content and thus was lacking a firm socialist content as noted by Tandon

(1985). In this way, "Defence Committees" were formed in work places and residential areas, all under the national umbrella of an ultra-leftist National Defence Committee. Despite all these "revolutionary" changes, the deteriorating economic situation continued to a degree that it became "an extreme case *even* in Africa" (*The Economist*, September 3, 1989, p. 40), with serious political repercussions.

From the perspective of the role of the human factor (HF) in Africa's development process, namely the positive or negative human characteristics and activities of people in the process of development, Ezeala-Harrison (1995, pp. 4–5) points out that a fundamental requirement for African development is:

> leaders resolved to honest and disciplined leadership, loyalty to their countries and people, totally committed to the development of their countries. . . . (They) must be willing to hand over political power to prospective successors rather than resorting to coups d'état, subversion, and treachery in order to perpetuate themselves in political power. They must eschew selfishness and greed, and thereby allow the process of orderly intergenerational succession to take a foothold in their countries.

This fundamental HF requirement was, however, not an issue at all for the PNDC government. In order to preserve its political power, at a time when it was most vulnerable to any form of manipulation by international capital, the government under the leadership of Rawlings adopted the metonymical policy of "economic rehabilitation for the production of wealth," under the auspices of none other than its "capitalist archenemy," international capital (the IMF and the World Bank). Ostensibly, the wealth subsequently produced could eventually be enjoyed by all Ghanaians, including the workers. In his radio and television broadcast to Ghanaians on May 2, 1983 regarding the newly adopted ERP, Rawlings asserted that it was "the only viable option open to us," and the only way to stop the downturn in the economy was to "stimulate production—particularly in the foreign exchange earning sector of the economy—by providing it with adequate financial incentives in the form of bonuses for foreign exchange earnings from legal exports." As we noted earlier, the "basic objective," according to Rawlings, was "to establish the right environment to do business and to make investments; both foreigners and our own citizens can then take advantage of the environment" (Ghana Investments Center, 1990, p. 7). In other words, it became cogent, expedient, and imperative to create an enabling economic environment for production, a move that entails political stability with Western support, and necessary for the operation of international capital. It was thus within the foregoing political and economic contexts that the ERP was initiated in 1983.

Much as the "basic objective" was laudable, because it was essentially a façade for the government's self-perpetuation, the program was not designed to simultaneously address both adjustment and structural transformation problems of the economy to make it relatively self-centered à la UN/ECA and in consistency with the tenets of the LPA to achieving spatial integration. This shortcoming is reflected in the formal wood-processing sector. It was, therefore, within the context of the adjustment metonym, that the World Bank (1984, p. xx) advised the PNDC government that "the forestry sector offers the greatest immediate potential for growth and foreign exchange earnings" at beginning of the program. In fact, the sector was particularly attractive and mutually convenient because it was seen as a sector with a readily available and vast natural resource base; capable of spinning off linkages within the domestic economy and generating foreign exchange to meet the nation's external commitments, while at the same time, satisfying the nation's role in the international division of labor as a producer of raw and semi-processed raw materials for the industrialized West. As a façade, the program gave the government a deep sense of purpose and urgency—an unflinching commitment to mobilize the forestry sector, along with other raw material production sectors such as mining and cocoa, to achieve its goal. In this regard, a centralized approach with respect to the forestry sector was thought to be most effective. Thus, the livelihood and interests of the forest communities were not part of the calculus of the government and international capital at all—at least initially (see Ryan, 2001). The rehabilitation of the forestry sector with adjustment loans constituted a convenient means for the Ghanaian government to satisfy the demands of international capital by keeping the country's hard currency revenues stable via wood and other primary exports. By so doing, the government could also keep itself in power, cushioned by the adjustment loans that the country would receive.

For international capital, rehabilitating the forestry sector was a priority because a large proportion of the lumber processed was exported as raw material in addition to logs for international consumption, prior to the deterioration of the sector. It is therefore not surprising that at the very beginning of the adjustment programme, the sector and its associated industries became the beneficiary of the second highest capital infusion (US $157 million) out of the World Bank loans from the government for rehabilitation (Ofei-Nkansah, 1991). Certain important policy measures under the program specific to industry in general, and the timber sector in particular had to be implemented to ensure the recapturing of Ghana as a production and trading turf for international capital, and a political turf for the Rawlings administration. In this regard, changes under the ERP regarding the structural and organizational patterns of the wood-processing industry become more meaningful in relation to the PNDC government's industrial policy under the program.

Industrial Policy under the ERP

The government's industrial policy under the program was couched in some related national development objectives (see Ghana Investments Center 1990, p. 17). These were:

a. the development of the production capacity of the economy through the efficient utilization of its resources and economic potential;
b. the full utilization and expansion of the productive capacity of existing enterprises;
c. the efficient saving of imports, the increase in exports, and the improvement of services which will assist in strengthening the balance of payments position of the country;
d. the encouragement of an even country-wide distribution of projects to encourage investments in the rural areas; and
e. the creation of a high level of employment and impartation of technical skills to Ghanaian citizens.

In connection with these national development objectives, the government issued a public statement in 1985 outlining the main objectives of its industrial policy and action program under the ERP, which was developed in close collaboration with the IMF and the World Bank teams which negotiated the Standby Agreements and the IDA and African Facility Credit from 1983 through 1986. The policy (see UNIDO, 1986, pp. 61–63), which sought to foster spatial integration among other things, was as follows:

a. to increase industrial output through greater use of existing installed capacity;
b. to improve supplies of essential goods to meet the integration demands of economic activities in the priority sectors of the economy;
c. to remove production bottlenecks in the most efficient industries and achieve significant cost reductions in others that are potentially viable through selective rehabilitation;
d. to develop economically viable linkages among local industries and between key economic sectors (e.g., agriculture, construction, transport);
e. to raise export earnings of non-traditional manufactured goods (as furniture and furniture parts).

In connection with these, the short term strategy under the program was to make available additional resources to the more efficient priority industries like wood-processing, which can make immediate supply response. The medium term strategy emphasized the rehabilitation of the potentially viable and efficient industries and the development of the local raw-material base for industries, taking into account the country's comparative advantage, in order to reduce the dependence on foreign inputs.

Lastly, the long term objective under the program was the achievement of an industrial structure that is more clearly linked to domestic inputs and needs that could provide a higher share of its own foreign exchange requirements. Such an industrial structure was also intended to sustain a rate of growth of output and employment above the economy-wide average but in balance with the rate of growth of resources and demand generated by agricultural development. As part of the long term objectives, non-viable excess capacity plants that constituted a drain on resources and could not supply the Ghanaian market competitively were to be phased out whilst continuing the process of rehabilitation in the most viable existing plants. Investments in new industries that could realize Ghana's comparative advantage and export potential were a priority under the policy. These long term objectives were meant to be supported by the ERP's policy environment in which success was based on efficient use of scarce resources and in which incentives are balanced between sectors and between growth and employment objectives. An action program, comprising six major policy measures, was also outlined for the pursuit of these industrial objectives and strategy. The measures, which were consistent with SAPs, were as follows (see UNIDO 1986, pp. 62–63):

1. Phased reform of production incentives to promote more efficient, competitive, resource-based, labor-intensive, and export-oriented industrial production, while giving firms time to adjust;
2. Greater access by efficient producers to imported industrial inputs and spare parts while progressively reducing administrative decisions and allocations;
3. Export incentive and promotion efforts to provide a return to exports that is equivalent to that on domestic sales and to compensate for disadvantages and costs that would reduce export competitiveness;
4. A positive climate for industrial operation and investment, including reasonable protection and investment incentives that are relatively uniformly and automatically available across broad classes of priority activities and that provide adequately for small-scale and employment-oriented activities;
5. Financial reforms to increase the ability of the financial sector to make credit available for industrial adjustment and rehabilitation, and to provide incentives for more labor-intensive techniques; and
6. More limited direct public sector participation aimed at making the most effective use of scarce public, financial, and managerial resources.

The government also took some policy actions to operationalize the policies, strategies, and program of action with respect to industry including wood-processing under the program. In order to improve the business environment for private enterprise and to establish an on-going rapport with the private sector, the government quickly inducted private sector representatives into policy-making bodies as the National Economic Commission and the Tripartite Committee which had the responsibility of dealing with prices, wages, and other issues involving government, management and labor. It also established an arrangement whereby bipartite meetings were to be held with such related groups as employers and associations such as the Ghana Timber Millers Organization to discuss matters of mutual concern.

In addition, the government created a wholly government-owned agency, the Non-Performing Assets Recovery Trust (NPART) with the function of recovering the non-performing assets of distressed banks. In specific relation to the formal wood-processing industry, insolvent firms were to be taken over by the agency and their debts paid by the Bank of Ghana, in the form of either stocks or bonds redeemable at a future date on the firm's behalf. NPART would then find a way to recover the money and this could be done by rescheduling the debt or selling it off. To facilitate the functions of NPART, a special judiciary tribunal was set up and empowered to speed up asset recovery processes. The centerpiece of the government's program to provide for greater participation of both domestic and foreign private enterprise in manufacturing, including wood-processing, was however, the enactment of the PNDC Law 116, the Investment Code of 1985, as a legal framework for the promotion of investments.

The Ghana Investment Code of 1985

This ERP-adapted investment code was designed to deal with investments in all sectors other than petroleum and mining. It was, in fact, a revision of the 1981 Investment Code (Act 437) to: (a) reflect government economic policies contained in the ERP; and (b), incorporate experience gained by the Ghana Investments Center in operating earlier legislation, as well as to reflect trends in investment policies of other developing countries with a history of successful investment promotion efforts (GIC, 1990, p. 17). This new investment code offered attractive special incentive packages to any entrepreneur who invested in the government declared priority areas such as wood-processing (e.g., in products such as sawn wood, dried lumber, floorings, furniture , and furniture components, for export), either as wholly foreign-owned, joint venture, or wholly Ghanaian enterprise. Under the code (see UNIDO, 1991, p. 51), the ventures were expected to contribute to spatial integration via

a. the development of the productive sectors of the national economy such as wood-processing under manufacturing;
b. the efficient utilization, expansion, and diversification of the productive capacity of existing enterprises;
c. utilization of local materials, supplies, and services;
d. the creation of employment opportunities in Ghana. They were also expected to contribute to
e. real increase in national export earnings;
f. real savings on national imports;
g. the development and transfer of advanced technology, including upgrading of indigenous technology; and
h. country-wide distribution of viable enterprises.

There were, however, certain stipulations for foreign participation and technology transfer. Under the code, wholly foreign owned ventures were to be net-foreign exchange earnings, and the minimum foreign equity was to be US $100,000. In cases of joint ventures with Ghanaian partner(s), the minimum foreign equity in investment must be US $60,000 and the structure of the company was expected to be determined by the partners. The code encouraged the joint-venture form as a means of ensuring that the balance of payments position is not burdened with outflows. It was also to ensure that the gains from investment and related benefits such as managerial, technical, and specialized know-how and skills are imparted to Ghanaian nationals.

The transfer of any industrial technology was to be in consonance with national development priorities in the sense that the technology was to be relevant and appropriate to Ghanaian conditions. As much as practicable such technology was to utilize Ghana's raw materials, supplies, and services. Also, the technology was to be proven and clearly defined in an agreement with the Ghana Investment Center (GIC). The duration of such agreements were not to be less than eighteen months, appropriate training programs were to be built into the agreements, and the payments for the technology were to be fair and reasonable. Finally, as far as practicable, restrictive clauses such as tied purchases, export restrictions, and transfer pricing were to be minimized. Under the code, transfer of technology agreements were acceptable only when the recipient enterprise is wholly Ghanaian-owned, or it is a joint venture between foreign and Ghanaian partners where the equity shareholding of the latter is at least 20%. In addition, the code provided general incentives, benefits, and concessions to entrepreneurs and investors, and special packages for specific sectors as manufacturing. The general investment incentives and benefits were designed to grant relief from taxation in various forms. Investors were granted full exemption customs import duties on plant, machinery, equipment, and accessories required for the enterprises and with accelerated depreciation. Manufacturing industries like wood-processing were given an investment allowance of 7.5% and depreciation or capital allowance of 40% in the year of investment and 20% in subsequent years.

To address the issue of regional disparities, the code included locational incentives in the form of tax reliefs. For wood-processing, there was an income tax rebate of up to 40% for locating outside Accra, but not in the traditional wood-processing areas of Kumasi and Sekondi-Takoradi, where the rebate is 25%. A deferment of income tax could also be enjoyed by firms locating in areas lacking basic infrastructure and where the firm undertakes the cost of such infrastructure (UNIDO, 1991, p. 54). Investors could enjoy the retention of a minimum of 35% of export earnings in an external account. For the wood-processing industry, the incentive is 20% and 35% of export earnings for lumber and value added products respectively held at the Ghana Commercial Bank, London, England, for purchasing equipment, spares, materials as well as for debt servicing, and payment of profits, dividends, and expatriate staff. The code also guaranteed immigrant quota in respect to approved expatriate personnel and an exemption from transfer tax for external currency transferred outside by expatriate staff. Firms were also exempted from the payment of the selective Alien Employment Tax for expatriate staff with approved quotas. In addition, they were guaranteed the transferability of profits, dividends, royalties, directors' fees, and capital in the event of sale or liquidation; they are also guaranteed against expropriation of the enterprise as well as against forced cession of any interest in the enterprise. As a protection for International capital, Ghana had to sign on to the Multilateral Investment Guarantee Agency (MIGA) convention initiated at the 1985 annual meeting of the World Bank. MIGA, as a protective mechanism, provides insurance cover for investors participating in eligible investments in the productive sectors of the country's economy. The insurance may be obtained against delays in currency transfers if such delays are attributable to the host government; expropriation and similar measures; breach of contract; and war and civil disturbances. In addition, Ghana had to sign on to Bilateral Investment Promotion and Protection Agreements (BIPA) with a number of countries to give further protection to investors from those countries.

To address the issue of unemployment or job creation for Ghanaians, in such situations where a manufacturing firm employed more than 100 Ghanaians, the company enjoyed an income tax reduction to the value of the social security contributions payable in respect of every Ghanaian employee in excess of the first one hundred. Regarding incentives on exports, manufacturers were entitled to a rebate on their tax liability for any year of assessment if, during that period, a percentage of the total output was exported: 5%–15% of total output exported attracted a rebate of 25%–30%; 15%–25% of the output attracted 33.3%–50% rebate; and finally, 25% or more of the output attracted rebates of 50%–60% on tax liability. In addition to the general incentives available to new investors, existing industries (i.e., firms registered before January, 1988), which reinvested at least US$ 100,000 or its equivalent using their profits or loans for expansion or rehabilitation enjoyed certain benefits under the code. They enjoyed an investment allowance of 10% in the year of investment, and were exempted from customs import duties and other related charges on plant machinery, equipment, and accessories imported exclusively to be used by the firm for its expansion or rehabilitation or both. They were also exempted from customs import duties and other related charges on spare parts not exceeding 100% of the value for the approved plant and machinery, to be used by the enterprise. It was in connection with all these, the government set up the Accra-based Ghana Investments Center (GIC), with a department devoted to wood-processing and charged it with the responsibility of promoting and encouraging investments in the economy under the ERP. It was also made responsible for investments approval under the program.

To promote and encourage investments under the ERP, the specific functions of the Center included (a) the collection, collation, analysis, and dissemination of information about investment opportunities and sources of investment capital. It also advises upon request, on the availability, choice of, or suitability of partners in joint-venture projects; (b) the initiation and organization of promotional activities such as exhibitions, conferences, and seminars for the stimulation of investments; (c) the identification of specific projects and the invitation of interested investors for the implementation of such projects. In addition, it was responsible for (d) granting approvals for the establishment of enterprises as specified under the code; (e) securing all licenses, authorities, approvals, and permits required to enable approvals granted by the center to have full effect; and (f) maintaining liaison between investors and ministries, government departments, agencies, institutional lenders, and other authorities concerned with investments. It was also charged with (g) monitoring and enforcing compliance with the terms and conditions of approvals granted under the code; and (h) approving and keeping records of all technology transfer agreements under the code.

Another important related government body that engaged in the promotion of the ERP with respect to the promotion of the export of processed-wood and other non-traditional items is the Ghana Export Promotion Council (GEPC). This organization had the responsibility of monitoring the improvement of the policy and administrative framework for exports. It was also designed to bring together the government and exporters, carrying out studies of export potential and problems and export promotion activities in Ghana and abroad. The specific responsibilities of the GEPC which also has a special section devoted to the wood-processing industry included: the development of national export awareness among potential exporters in the manufacturing industries and other sectors, and helping in the identification of the products and locating markets for them. The GEPC also creates goodwill for such products through local and overseas trade fairs and exhibitions and provides the exporters with the necessary assistance for penetrating competitive international markets.

The council was also charged with training exporters to upgrade their skills in export marketing. In this respect, it established the Export School to develop training programs covering production, processing, marketing, packaging, procedures, documentation, and all other aspects of the export function. The training programs were delivered through seminars and workshops using both local and foreign experts in a wide range of export-related issues. It also recommended to the government the necessary assistance and incentives needed by the Ghanaian exporter. Since 1990, GEPC has required exporters to pay an annual registration fee of ¢25,000.00. Responses from the wood-processing and other sectors of industry to the incentives and benefits administered by both the GIC and GEPC were for the most part, influenced by the government's exchange rate policy and reform under the ERP.

Exchange Rate Policy and Reform under the ERP

Exchange rate policy and reform were a critical part of the ERP, particularly with respect to redirecting incentives toward productive activities and exports, and to endowing the nation with an efficient exchange rate system. In the pursuit of these, the government started the devaluation of the *Cedi* (¢) in 1983 and periodically thereafter, to gradually attain its true value vis à vis the US dollar and other major currencies. At the beginning of the ERP in April, 1983, the exchange rate was pegged at ¢2.75 to US $1.00. By December, 1990, it was ¢345.00 to the dollar, and ¢358.80 to the dollar by May, 1991.

An auction system was also established initially with a view to allowing the *Cedi*'s exchange rate to be determined by market forces. The auction system allowed the central bank, Bank of Ghana, to continue to centralize the nation's foreign exchange receipts, which was viewed as crucial in minimizing capital flight. It was also a preferred mechanism because at least in principle, it could prevent the emergence of collusive behavior on the part of the commercial banks. The legalization of the foreign exchange (forex) bureaux was also a measure to absorb the parallel market for foreign exchange (Kapur et al, 1991, p. 19). As a reflection of the retreat of state activism with respect to the currency market and to allow the "invisible hand" of the market to operate, any bank, person, or institution licensed by the Bank of Ghana could operate a forex bureau. Operators were authorized to engage in the purchase or sale of foreign exchange at freely negotiated rates and each bureau was free to quote buying and selling rates. Subject to certain rules issued by the Bank of Ghana, the forex bureaux were not required to identify their sources of foreign exchange; likewise, sellers of foreign currency to the forex bureaux were not required to identify their sources either—thus broadening the transaction base of the exchange forex bureaux and facilitating the absorption of the parallel market. Consequently, as noted by the Kapur et al. (1991, p. 21), on the demand side of the new market, all bona fide import and service payments could be funded through the forex bureaux. In addition, capital transactions, although illegal in certain cases since controls continue to be applied, were a source of demand for foreign exchange in this market. On the supply side, the key sources are foreign exchange receipts from non-traditional exports, private remittances, and foreign exchange held by the public.

On April 27, 1990 however, the Bank of Ghana began a wholesale foreign exchange auction and discontinued the retail auction. Any licensed forex bureaux could be eligible, if they did not bid on their own account; and with the marginal exception of bids covering the dealers' import needs, dealers' bids were to be entirely backed by firm bids by end-users of foreign exchange (Kapur et al., 1991, p. 23). All authorized dealer banks were also declared eligible to participate in the exchange auction. This development resulted essentially in full liberalization of the exchange system, allowing the achievement of significant progress toward the restoration of financial convertibility of the *Cedi*, as there were no longer any government limitations on the making of payments and transfers for current international transactions (Kapur et al., 1991, p. 24).

There were other general stabilization and structural measures that affected the wood-processing industry (GIC, 1990, pp. 14–15, p. 34). Personal tax, corporate tax, and excise duties were rationalized in order to simplify the tax structure with a continuous review of tax rates (corporate and personal) in order to make taxes competitive to private investors. For example, corporate tax declined from 60% of taxable payment to 45% for industrial prices. Also, for most dividend earners, there was a decrease in tax rate from 55% to 30%. The interest rate (and hence borrowing and lending rates of capital) was freed from administrative determination. In fact, since October 1987, interest rates have been market-determined on the basis of a discount rate derived from a weekly Treasury Bill Tender System. Also monetary policy since 1985 was aimed at restricting the growth of money supply and hence, to reduce the rate of inflation. Consequently, the inflation rate declined from 123% to about 32% in 1988, and fell further to about 25% in 1989 (compared to the planned 15% rate) and by May,

1991, it was 33.3% (Kapur et al., 1991, p. 52). Even though availability of credit was controlled to keep inflation down, credit expansion was allowed in 1987 through 1989 just to ensure adequate financial support for the priority sectors of the economy. In addition, the bureaucratic import licensing system was abolished in order to facilitate the imports of goods needed for domestic production, particularly raw materials (UNIDO, 1991, p. 60). Furthermore, measures were introduced, under the on-going state-owned enterprises (SOEs) reform and divestiture program, to improve the performance of the SOEs.

Lastly, the country's infrastructure (roads, railways, ports, telecommunications network, educational and training facilities) were rehabilitated. Since the economy was highly dependent on the production of export commodities as timber and processed wood, some 40% of the import program in 1984 was allocated to transportation and petroleum sectors. Four World Bank projects focused on transport infrastructure. The rehabilitation of the railways was supported by the World Bank/African Development Bank's Reconstruction Import Credit and the government funds the maintenance work on the railways on priority basis. In 1986 for example, Fr 80 million out of a Fr 100 million loan from France for the industrial sector was allocated to the transport, telecommunications, and electricity sectors. Japan, Saudi Arabia, and the World Bank also provided loan finance for port development at Tema and Takoradi. The ports were dredged to enable them to handle larger vessels, the wharves rehabilitated, quays strengthened, and new floating craft and cranes acquired. There has also been the construction of a container handling platform by January 1992 to meet the hitherto unavailable modern day container traffic (see the January 6, 1992 issue of the *People's Daily Graphic*, #12787, pp. 1 and 9). In connection with this, the government adopted a payment system at the ports which demanded payment in foreign exchange for all services performed for the ship-owners. This was to enable the Port Authority to use some of this revenue to maintain the port equipment (GOG, 1984, p. 13).

In all, the process of import liberalization was completed on February 5, 1988 and effective January 14, 1989; the Import Licensing System was abolished (Kapur et al., 1991, p. 20); and the daily minimum wage was raised on March 28, 1990, to ¢218.00 as a reflection of the government's efforts to safeguard a minimum standard of living for the unskilled segment of the labor force (Kapur et al., 1991, p. 53). The price of premium gasoline per liter was adjusted from ¢5.50 in April 1983 to ¢200.00 by the end of March, 1991 (Kapur et al., 1991, p. 54).

The focus of this chapter has been on the political economy of the ERP and the formal wood-processing industry. In connection with that, the Ghanaian political economy was situated in a global context to reiterate the argument previously made, that the problematic patterns of the industry were largely a legacy of Ghana's colonial history. The political and economic developments that culminated in the adoption of the ERP was discussed, integrating the locus of the wood-processing industry in the process. The central argument was made to the effect that the ERP was essentially metonymical in that, at the time of its adoption, it constituted a convenient strategy to defuse two major problems that confronted the PNDC government by 1983. These were the domestic political and economic crisis; and the certain collapse of the government if it failed to accommodate international capital as evidenced by the fall of two previous governments that had confronted international capital. In respect to these, the liberal and concessionary nature of the program's macroeconomic policy measures, in terms of the strategic combinations of demand management and structural policies, aimed at mitigating pre-ERP problems and fashioning a conducive environment business-wise, for industry and other sectors of the economy including wood-processing were presented. In the next chapter, we look at policy specifically geared at the timber industry and the implications of the program for trade and for the environment.

Chapter Five

Turf Recapture and Consolidation for the Timber Trade

Ghana's Economic Recovery Program

By 1983, international capital had virtually lost the Ghanaian economy as part of its turf, and the PNDC government, after its attempt to delink by refusing to accommodate international capital by its socialist stance, was also losing its control on its political turf. These losses of turf control were demonstrated in the state of the economy in the 1970s through 1983. Recounting the trend, the Ghanaian economy experienced a precipitous decline, beginning in the latter part of the 1970s and reaching its lowest level in early 1983. This decline was vividly demonstrated in the forestry sector, an accumulation turf for international capital, which had traditionally been the third most important component of the economy because of its contribution to the country's foreign income generation via tropical wood exports. The government was losing its legitimacy due to the deteriorating state of the Ghanaian economy, so at the beginning of 1983 its only viable option was the adoption of a structural adjustment program as a strategy to resuscitate the economy, and thus entering into a "Faustian bargain"—a deal contracted for present gain without regard for future cost or consequences—with international capital. Under the program, the nature and economic contribution of the forestry sector provided a satisfying condition for it to be exploited, as a matter of mutual immediate convenience for Ghana's PNDC government and international capital. This was because of the sector's potentially rapid income-generating nature—a "goose" that could quickly lay "golden eggs" for the benefit of both parties.

Along with the broad government industrial policy couched in the national development objectives under Ghana's adjustment program, which emphasized spatial integration, as discussed in the last chapter, specific policies and actions were employed to develop a viable and efficient timber industry right from its roots, the forestry sector. Government policy under the ERP with respect to the forestry sector was to ensure improvement in the timber sector's deteriorating facilities, proper husbandry and efficient utilization of the country's forest resources (Forestry Commission, 1988, p. iv). Thus, the government's strategy for the recovery and rehabilitation of the industry was to increase the productivity of the forest base to the maximum, consistent with the need to protect the environment; a measure which includes the protection of watersheds and forest preservation (GOG, 1984, p. 11).

With specific regards to the timber or wood-processing industry, with the aim of "achieving a quick increase in production of the private companies through providing them with spare parts, equipment and materials," the government's policy was to support the rehabilitation of selected mills, whilst re-organizing state-owned ones (GOG, 1984, p. 11).

In line with that aim, the government moved to alleviate the immediate related physical constraints in the form of production equipment shortages, transport equipment, and spare parts with the assistance of sector adjustment loans (SECALs) from the International Development Association (IDA) to fund the export rehabilitation project. There were other SECALs from such international organizations as the Overseas Development Administration (ODA) and the Canadian International Development Agency (CIDA). The IDA, for example allocated US$ 28 million as export rehabilitation credit to the industry during 1985–1986 as a soft loan to the Government of Ghana. This SECAL had a repayment period of 40 years including a 10 year grace period with an average interest rate of about 1.5% per annum. However, the beneficiaries were to start repayment within 12 months of receiving the facility and were all supposed to pay up fully by 1989. The United Kingdom, to keep and sustain its trading turf, provided the Ghana government with some £8.7 million in 1984 and 1985 through the ODA as a grant tied to the purchase of timber sector equipment from U.K. manufacturers. In addition, the Export Credit Guarantee Department (ECGD) of the U.K. insured commercial bank credit with hard currency repayment guaranteed by the Bank of Ghana, to import a variety of U.K. sourced vehicles and logging equipment in 1983–1984, with a total credit of £7.6 million. The Export Rehabilitation Technical Assistance Project also assisted in the implementation of the industrial reforms and rehabilitation (Forestry Commission, 1988, pp. 2–6). In fact, some of the SECAL facilities were offered a food aid program, such as free vegetable oil and corn meal for the employees of the beneficiary companies to make the loans more attractive, at a time when hunger was acute in the country as part of the economic malaise (see Owusu, 2001a).

To ensure a steady flow of logs for processing (and for export in the round), 54.4% of the credit went to logging equipment with 35% concentrated on integrated mills; the majority of logging companies and individuals therefore did not benefit from that. The few beneficiary companies however could significantly increase their output, driven by the need to repay the loans. Various aspects of wood-processing absorbed 45.6% of the loan with no benefits going to the support services through which control and regulatory measures are implemented. The total infusion of US$ 157 million allocated for the rehabilitation of the forestry sector including the wood-processing industry, included the allocation for resource management. Specifically, the credit facilities were disbursed under eight beneficiary projects as shown in table 5.1.[1]

Wood-processing, including logging equipment for integrated mills, enjoyed 70.2% of that SECAL facility while the remaining 18.8% went directly into logging. The disbursement for logging equipment for integrated mills was used for capacity utilization improvement in the processing of such export items as lumber, moldings, and veneer. Part of it was also used to replace bush and transport equipment to suit the mills' intake requirements. It was also partly used to boost the financing capabilities of the seventeen beneficiary companies. The Logging Equipment for Concessionaires Project was aimed at improving the general wood supply situation and the activation of idle concessions. In connection with these, logging and transport equipment were provided in consistency with their respective financing capabilities and concession sizes. The Logging Equipment for Log Exporters Project was for log exporting companies interested in purchasing transport and logging equipment. It was judged that such investments would be efficient in terms of export generation based on their previous performance.

In the case of the Sawmills Improvement Project, the nine beneficiary established mills used the credit for the acquisition of more accurate head rigs and to boost efficiency and output through the purchase of drying kilns, molding, and transport equipment, usually under ODA/ECGD finance, with

Table 5.1. Distribution of Credits by Project Type

Project Type	*Number of Companies*	*Credit (US$)*	*% of Total Credit*
Logging equipment for Integrated Mills	17	18,855,500	35.6
Logging equipment for Concessionaires	29	8,878,000	16.8
Logging equipment for Exporters	4	1,060,000	2.0
Sawmill Improvements	9	15,807,500	29.9
Wood Energy	1	190,000	0.4
Furniture Exports	13	4,700,000	8.9
New Sawmills	5	2,510,000	4.7
Wood Preservation	1	928,500	1.8
Total	79	52,929,500	100.0

Source: Alder (1987) in Forestry Commission (1988, 17).

IDA funds used for processing machinery. The purchase of drying kilns was part of the program to phase out the production of air-dried lumber by 1994 since kiln-dried wood is more valuable and durable. Kiln-drying saves more production time and is thus faster and economically more efficient than air drying. As regards the Wood Energy Project, the funds were channeled into the production of sawdust briquettes as a substitute for fuel wood for urban industrial use such as in bakeries. Only one company received this facility because it was thought that export prospects of the briquettes did not appear good. The Furniture Exports Project was geared at a wide range of exports. It included furniture parts, knock-down furniture kits, and modular panel products such as wardrobes, cupboards, as well as fixtures such as solid and flush doors. Inclusion of firms in this category was justified on the basis of export contracts they held. The New Sawmills Project involved five new companies which presented proposals involving concessions, logging, and sawmilling, one of which included a veneer peeling line. Lastly, funds for the Wood Preservation Project went to only one company for the development of a pressure impregnation plant with kiln-drying equipment for the production of treated transmission poles for export to other developing countries. Table 5.2 is a summary of imports under the IDA's 1985 and 1986 program.

On the whole, the IDA credits for the 1985/1986 program, the largest among the three, benefited sixty firms. To make this possible, local capital in the form of nine local banks namely, National Investment Bank, Social Security Bank, Ghana Commercial Bank, Barclays Bank of Ghana, National Savings and Credit Bank, Standard Chartered Bank (Ghana) Ltd., Merchant Bank, Bank for Housing and Construction, and Agricultural Development Bank offered guarantees, in various degrees to the sixty companies (see Boateng, 1991, p. B-63). Two extra companies had theirs guaranteed by the Ministry of Finance and Economic Planning. Out of the total of US $28,842,000.00, US $17,688.00 went into the purchase of equipment; US $1,682,000.00 for the purchase of materials; and US $9,472,000.00 was absorbed by the purchase of vehicles (Boateng, 1991, p. B-65). In addition to the SECALs channeled to the sector for rehabilitation and expansion under the ERP, the government undertook organizational reform and institutional strengthening at the behest of International capital.

ORGANIZATIONAL AND INSTITUTIONAL REFORMS AFFECTING FORMAL WOOD-PROCESSING UNDER THE ERP

We noted in chapter 3 that the bureaucracy associated with the GTMB, as the sole government agency charged with the overall organization of the industry, and its monopsonistic status established a pattern of inefficiency,

Table 5.2. Total Imports under the 1985/86 Programs (IDA)

Plant & Transport (Units)	*Quantity*
Logging Bulldozers	24
Rubber-tired Log Skidders	22
Log Loaders with Forks	9
Log Loading Winches & Anchor Tractors	15
Road Graders	2
Forklift Trucks	15
Heavy Trucks (12 Tons and up)	161
Light Transport, 4WD, Pickups	47
Processing Machinery/Equipment	*Credit (US $'000s)*
Sawmilling Machinery	2,040
Furniture & Molding Machines	2,993
Veneer & Plywood Machinery	805
Kiln Dryers	448
TOTAL	6,286
Spares	*Credit (US $'000s)*
Spare Parts for Mobile Plants	446
Spare Parts for Transport Equipment	511
Spare Parts for Processing Machinery	2,040
TOTAL	2,997
Glues, Lacquers, Preservatives, Insecticides, etc.	1,682

Source of Data: Boateng (1991, B-63).

corruption, and mismanagement that affected the overall organization and operations of the industry (see Killick, 1978). The activities of the GTMB thus became inimical to the growth of the timber industry, in particular, and the nation's development in general, a deterioration that festered especially after 1966 and reached its peak by 1983 when the free-market based ERP was adopted. As part of the organizational and institutional reform affecting the timber industry under the ERP, the moribund GTMB was finally dissolved in 1986. Its duties were taken over by the Timber Export Development Board (TEDB), and the Forest Products Inspection Bureau (FPIB), each with specific specialized functions, to ensure more efficiency. Let us now look at the specific functions of each of these two bodies.

The Timber Export Development Board

This body was established under the PNDC TEDB Law 123, 1985, and charged with the responsibility to promote the sale and export of Ghana timber products; develop markets for, and promote the sale of lesser known timber species; and collate, analyze, and disseminate relevant market intelligence and ensure its frequent publication. It was also required to make projections and forecasts of market demand trends for timber and timber products; and to compile and publish timber export statistics in cooperation with the Forest Products Inspection Bureau (FPIB) established under the FPIB Law, 1985 (PNDC Law 117). In addition, it was expected to liaise with timber marketing associations and main timber producers for the rapid development and servicing of export markets; and publish journals and periodicals reporting the results of market surveys, current demand for species, quality and prices of all overseas buyers for the benefit of the timber trade, as well as do all such other things as are incidental to the foregoing functions. Lastly, in line with the government's policy to avoid overextending itself, the TEDB was expected to perform these functions on sound commercial lines.

To meet its obligations to the timber industry, TEDB devised an export marketing strategy as shown in figure 5.1. According to this strategy, which

Figure 5.1. TEDB Export Marketing Strategy
Source: Adapted from TEDB (1991)

addressed both the supply and demand issues pertinent to the market, all registered exporters of Ghana's timber—processed wood and round logs—had to forward their wood availabilities to the board at Takoradi. On the supply side, TEDB then relayed the information electronically—either by telex or fax—to their United Kingdom office in London. The London office would then make the information available to its buyers who can sell anywhere in the world to consumers, importers, or manufacturers.

In 1991, for example, there were 84 registered buyers. For regular and planed lumber the buyers were as follows: 1 each from Canada, Malta, and Portugal; 2 each from Denmark, Holland, Spain, and Switzerland; 3 each from France and Belgium; 4 from the Republic of Ireland; 7 from the USA; and 29 from the U.K. For furniture components/knock-down furniture, there were 2 and 8 registered buyers from the Republic of Ireland and the U.K respectively, and 1 each from Canada, Germany, Holland, Portugal, and Sweden. On the demand side, the needs of the consumers, importers, and manufacturers outside Ghana had to be channeled through TEDB's overseas buyers to the board's London office. The London office then relayed the demand to Takoradi from where the Ghanaian suppliers were notified of the demand.

One of the incidental actions of the board is the establishment of the only saw-doctoring school in Sub-Saharan Africa to cater to one of the technical needs of the industry, and has plans of setting up a Timber School. Also, the TEDB, Ghana Furniture Producers Association (GFPA), the Ghana Educational Service (GES), the Ministry of Finance and Economic Planning, in conjunction with the European Economic Commission, commissioned an ECU 150,000 rehabilitated model woodworking shop for Accra Polytechnic. The workshop was equipped with such new equipment as Tenoning-jigs and routers as well as saw sharpening and dust extraction machines.[2] It thus served as a model wood-processing factory as well as a training ground for personnel in the wood industry. The training emphasized export furniture designing, production, and management control. The TEDB also became very instrumental in the organization of the periodic international exhibitions which attract buyers, sellers, and investors worldwide. Whilst the TEDB deals with export promotion, production development, and marketing research, the FPIB was designed to concentrate on contracts, approvals, and quality control.

The Forest Products Inspection Bureau

This body was also established with the government's FPIB Law, 1985 (PNDC Law 117), and charged to recommend to the PNDC Secretary responsible for Lands and Natural Resources, timber grading rules for standard use in Ghana, and to disseminate timber grading rules prescribed by the Secretary for standard use in Ghana. The bureau was also charged to ensure that the established grading rules were applied by the timber industry, and also to recommend to the Secretary regulations and procedures for the stamping, tagging, and marking of logs and timber products. In addition, it had to undertake the actual stamping, tagging, and marking of logs and timber products, as well as the training, examination, and registration of authorized timber graders and establish levels of certificate for such graders. The other requirements were for the bureau to conduct pre-shipment examination of logs and timber products for export to ensure correct grading and valuation and to eliminate fraud, and to vet and register all contracts for the exportation of logs and timber products. The bureau also had to register claims in respect to contracts for the exportation of logs and timber products and to appoint arbitrators in disputes involving grading and valuation of timber products, register all enterprises engaged in the timber and wood-processing industries, and lastly, was required to do all such things deemed reasonably incidental to the mandated functions. In this regard, it was expected to halt, open, and detain for purposes of inspection, any vehicles, container, or vessels carrying or containing or believed to be carrying or containing timber or wood-products.

As part of their requirement to operate on sound commercial lines, both the TEDB and the FPIB had to levy charges of 1% and 2% respectively on the f.o.b. value of wood exports. In addition to the registration fees paid to the GEPC, the exporters also register with FPIB at a fee of ¢30,000.00 as well as with Ghana Shippers Council at an annual fee of ¢50,000.00 to ¢150,000.00 depending on the exporter's capacity. To be an exporter of processed-wood, membership in the Ghana Timber Association (GTA) and/or Ghana Timber Millers' Organization (GTMO) is mandatory.[3] Members also contribute 1% of their respective incomes from all exported mill and log timber to finance the nation's re-forestation programs.

Organization of Processed Wood Exports under the ERP

A number of steps were also developed with respect to the inspection of the exports, documentation, and shipping, as part of the quest to eliminate fraud (see TEDB Export Guidelines and Boateng, 1991, pp. 16–17). This was done by requiring FPIB inspectors to undertake quality, quantity and dimension specification inspection, when the processed wood products are ready at the factory or mill. They would then issue the necessary certificates, with frequent re-inspection carried out at the port. The exporter then had to complete a Bank of Ghana Foreign Exchange Control Form (Form A2) from the bank handling the letter of credit (L/C). The FPIB inspection certificates, the A2 form, the L/C, and the exporter's contract could then be used to apply for an export permit from FPIB. Upon the issue of the permit, a Pre-shipment Notification Form (costing ¢1,500.00) and a Shipment Notification Form (costing ¢5,000.00) had to be obtained from the Shippers Council and completed, and then all the documents examined and stamped by a Ministry of Trade representative, after which the exporter proceeded to the Customs for numbering and checking. All the documents were then to be presented to the port's PNDC Security Coordinator (seconded from the Ghana Armed Forces) to be signed and stamped, after which the harbor office of Customs "released" the cargo. The release authorized the Port Authority to allow the cargo onto the vessel.

Finally, after the loading and the issuance of bills of lading, the exporter could then proceed to produce invoices and other papers for the bank to release export proceeds. In addition to all these steps, the exporters or their forwarding agents had to lodge a bond, usually in the form of a bank guarantee of ¢4 million with Customs. These measures were all taken to ensure maximum returns from timber exports. The quest to maximize returns from the export of timber however cannot be divorced from the environmental implications of the extraction, namely the resource base—the forest.

THE TIMBER INDUSTRY AND THE ENVIRONMENT UNDER THE ERP

Apparently conscious of the environmental implications of the quest to maximize returns from timber exports under the program, Mr. P.V. Obeng, PNDC member and Chairman of the Committee of Secretaries remarked in London in 1988 that:

> Some observers of Third World development have been obliged by experience to question the wisdom of large scale use of natural resources in pursuit of economic goals. Let me assure all such observers that we in Ghana recognize the necessity to maintain a sustained and sustainable yield. We are also very much aware of the ecological consequences, both to ourselves and to the living Earth, of the destruction of tropical forests.[4]

From the perspective of the government, the resource base was therefore receiving particular attention under the program as was reflected in a forest policy that was developed in that respect. As indicated in the TEDB's document, *Basic Information, Facts and Figures* (TEDB, 1990), the key aspects of the Forest Policy framework under the ERP for the resource base entailed the creation of a permanent forest estate by reserving appropriate areas of forests or land suitable for reforestation and forestation; and the management of a permanent forest estate to achieve maximum productivity and value in perpetuity. In addition, it required safeguarding water supplies and soil fertility, minimization of damage by floods and erosion to rivers and agricultural land, as well as the conservation of adequate forest areas for recreation, education, research, and protection of the country's flora and fauna. Furthermore, the domestic markets for the forest products were to be developed and the timber industry's production base was to be strengthened and diversified to eliminate waste and production problems. In connection with the policy, seven more species of wood were added to fourteen other species which had been banned from export in the log form, since 1988. These now were all to be processed to boost the domestic manufacturing industry and to increase their export values as part of the quest to derive maximum benefits from the resource base.

In order to manage Ghana's forests on a sustained yield basis under the program, a multi-donor mission of ODA, The Danish International Development Agency of Denmark (DANIDA), The Food and Agriculture Organization (FAO), and the World Bank reviewed the forestry sector under the Global Tropical Forest Action Plan in 1986. The study resulted in the 1989 implementation of the US$ 64.5 million Ghana Forest Resources Management Project (GFRMP), which was scheduled to last six years. This project involved forest management, sector policy reforms, rural forestry, institutional strengthening, forestry education, and training as well as adaptive research. Contributions from the government of Ghana, IDA, ODA, and DANIDA were US$ 9.7 million, US$ 39.4 million, US$ 7.1 million, and US$ 8.3 million respectively. In connection with the wood-processing industry, under the GFRMP, there were technical and financial assessments of all timber processing mills in 1991 to provide a basis for classifying the mills and firms into operational groups, each of which will receive appropriate incentive packages (Ofei-Nkansah, 1991, p. 229). In order to ensure that the proverbial "goose"—the forest—continued to lay the "golden eggs," within the colonial context, the creation of the enabling environment was done with due cognizance of the resource base, the forest, and its sustenance.

ADJUSTMENT-RELATED CHANGES IN THE INDUSTRY

The economic changes following the use of structural adjustment policy measures, as a mutual strategy by both international capital and the government to recapture their respective turfs were dramatic. In fact the adjustment-related changes were touted by the sponsors and the international financial community as a big success. *The Financial Times* of London's issue of July 20, 1989, in a special supplement on Ghana, for example, described Ghana as a "Star Pupil" of the structural adjustment process—a claim that is widely disputed, particularly by critics of the IMF and the World Bank, who see it as evidence of an accommodation of international capital.

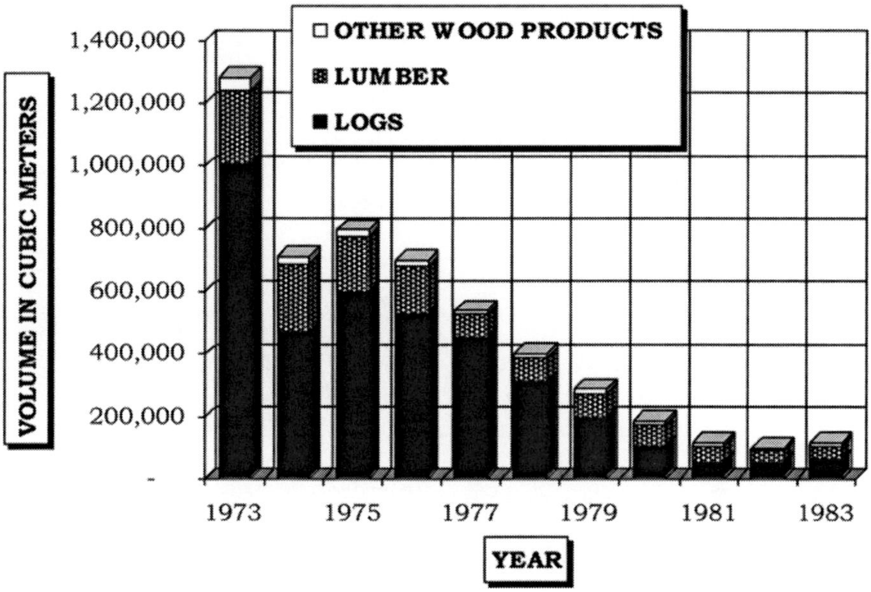

Figure 5.2. Volume of Timber & Wood Products Exported (1973–1983)

The apparent success was reflected in the timber industry with an adjustment-related expansion in the industry's export output. In the vested interest of the sponsors, namely international capital, this turnaround in the industry's export output continued to be dominated by logs, the basic raw material for the wood processing industries in Europe. This was followed by lumber, which in essence is also a raw material—with a little added value, vis à vis downstream products like furniture parts, plywood, and veneer production. To put these in changes in perspective, in 1973, ten years prior to adjustment, the forestry sector exported a combined total of about 1.3 million m^3 of processed and unprocessed wood, which fetched the Ghanaian economy US $136 million in foreign exchange. By 1982, the volume of export and revenue figures had systematically declined to 100,303 m^3 and $15.3 million respectively. The trend is shown in figure 5.2. In 1973, log exports, the dominant component, constituted as much as 78% of the total exports to Europe as compared with 19% for lumber and 3% for other wood products. In terms of volume, 1,006,848 m^3 of logs, 234,829 m^3 of lumber, and 41,049 m^3 of other wood products were exported that year. Even in 1982, when the industry was at its lowest ebb, the export proportions for logs, lumber, and other wood-products were 53%, 40%, and 7% respectively.

As shown in figure 5.3, however, a dramatic turn-around began to take place in 1983, the year the adjustment program began with a steady expansion

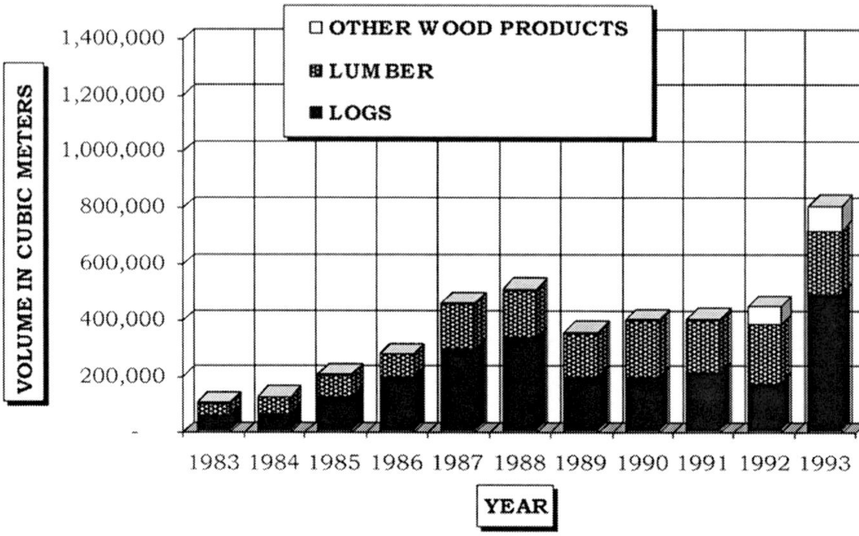

Figure 5.3. Volume of Timber & Wood Products Exported (1983–1993)

in timber exports. As in the pre-adjustment years, the dominant export continued to be logs, the most preferred or primary raw material among European wood industries. This was followed by lumber, which scored a 503% increase above the 1983 figure.

Downstream products like knock-down furniture and veneer also recorded very significant increases. The increases in the export of processed wood as shown in table 5.3 also gave semblance of increased manufacturing/processing activities in the industry consistent with the government's industrial

Table 5.3. Volume of Processed Wood Exported, 1983–1993 (M^3)

Year	*Lumber*	*Other Processed Wood*
1983	44,752	1,942
1984	56,409	2,156
1985	76,406	2,228
1986	83,577	699
1987	161,927	3,440
1988	169,640	5,637
1989	154,033	5,151
1990	200,517	5,595
1991	182,644	8,304
1992	211,427	63,383
1993	224,914	89,755

Sources of Data: Timber Exports Development Board, Ghana; Forest Products Inspection Board, Ghana; Ministry of Lands and Natural Resources, Ghana.

policy, particularly in promoting *more efficient, competitive, resource-based, labor-intensive and export-oriented industrial production.*

The expansion of export-oriented output is also reflected in the rate of capacity utilization of the installed equipment, which shows a tremendous increase after 1984 (CBS June 1983, 10). In the pre-adjustment years, as shown in table 5.4, the rates for all manufacturing and the wood processing sector had fallen steadily from 40.4% and 36%, respectively, in 1978 to 21.0% and 20.0% respectively in 1982. However, the rate for wood-processing began to rise steadily after the adjustment program was initiated in 1983, through 1987, and then stabilized at 70% for three years and 65% thereafter, relative to figures below 46% for the manufacturing sector as a whole over the period.

Significantly, the major increases in the rate of capacity utilization in the wood processing sector appeared after 1986, after new investment in the form of fresh equipment and machinery imported during the 1985–1986 programs had been installed. Despite the significant increases after 1983, the firms still had not been able to attain their respective 1978 production levels.

The steady increase in the rates of capacity utilization within the industry (and in fact, all manufacturing) after 1983 suggests that changes in the macro-economic conditions under the ERP were significant in the operations of wood-processing firms (and the manufacturing sector in general). Let us now examine the extent to which certain factors associated with the program, as

Table 5.4. Estimated Rates of Capacity Utilization: Large & Medium Scale Factories (%) 1978–1993

Year	*Wood Processing Sector*	*All Manufacturing Industries*
1978	36.0	40.4
1979	36.4	33.1
1980	27.3	25.5
1981	n/a	n/a
1982	20.0	21.0
1983	20.0	30.0
1984	28.1	18.0
1985	32.5	25.0
1986	n/a	25.0
1987	43.0	35.0
1988	70.0	40.0
1989	70.0	40.6
1990	70.0	39.8
1991	65.0	40.5
1992	65.0	44.5
1993	65.0	45.7

Source of Data: Statistical Service. *Quarterly Digest of Statistics*, June 1989a (p.10) and March 1995 (p.12); Central Bureau of Statistics. Economic Survey, 1982 (p. 6).

per Owusu (2001b), promoted increases in production (henceforth "enabling factors") in the timber industry and the aspects of the program that fettered the potential for an even greater output (henceforth "constraining factors"). For our discussion purposes, both sets of factors determining output for export are each loosely decomposed further into, and discussed under three categories relating to: 1. Capital; 2. Labor; and 3. Institutional and other factors, bearing in mind that these categories are, nevertheless, not mutually exclusive.

ADJUSTMENT FACTORS DETERMINING THE INDUSTRY'S EXPORT-ORIENTED OUTPUT

Capital-Related Enabling Factors

Capital investment in the form of SECALs such as the IDA, ODA, and ECGD credits totalling US $52,929,500, infused into the wood processing industry gave it a much needed booster, given the financial problems the firms had during the economic malaise, which was reflected in such problems as broken and obsolete equipment. Access to new equipment and spare parts using the new financial resources were tied to the contribution of the standardization of equipment, to boost output for export (see table 5.2). In this respect, the infusion of SECALs into the industry was useful to a limited degree. As part of the tied-purchases associated with some of the loans, some machines and equipment necessarily had to be imported from the countries that provided those loans as part of the agreement. Such mandatory imports ensured some degree of standardization, thus making synchronization in production for certain firms much easier and resulting in higher productivity.

As shown in figure 5.2, Ghana's wood had traditionally been exported mostly in the form of unprocessed logs. In order to generate more income from the forestry sector, the government instituted measures to augment the export of the higher-valued processed wood, and phase out log exports by 1997. In terms of capital investment, lumber production provided the next least expensive way of increasing exports with a ready external market. Processing the logs into lumber for export in huge quantities thus replaced the gradual cutback in log exports to make up for the loss of income in the latter activity. As part of the creation of an enabling environment for investments by local and foreign capital under the program, the wood processing firms were given incentive packages under the investment code to encourage exports—the more a firm exported, the greater the benefits it enjoyed. Consequently, those provisions discouraged the sale of lumber on the domestic market in preference for the more lucrative exports (Owusu 2001b). Specifically, the formal sector's sawmillers preferred to enjoy the tax rebates based on volumes exported, and could retain as much as 35% of the export earnings in an external account held at the Ghana Commercial Bank, London, England. In addition, they were permitted to use their export retention funds to meet business promotion, training, and related expenses; for debt servicing and payment of profits, dividends, and expatriate staff; as well as for the purchase of machinery, equipment, and spare parts and other materials to support further exports (World Bank 1991, p.82).

The exported lumber and furniture parts constituted the bulk of the post-adjustment expansion in processed wood exports. As we noted earlier, however, these are essentially raw materials for firms in the metropolitan or donor countries, and expansion in such exports is thus tied to the accommodation of international capital. Consequently, the imperative of meeting the needs (accommodation) of international capital by the Ghanaian government was a major determinant in the increased exports of processed wood. This accommodation was clearly demonstrated by the fact that the bulk of the increased exports went to the donor countries, mostly in the then European Community (EC), now the European Union (EU). For example, from 1983 through 1991, a minimum of 81% of the volume of processed wood exported was sent to EC countries annually. Together with supplies from other Sub-Saharan African tropical wood-producing countries, Ghana's processed wood exports became so inexpensive that resale often took place on the world market (see Barker et al. 1991, p. 5). Such a glut on the international market thus supported the critics' argument that SAPs, *inter alia*, provide cheap and reliable sources of raw materials for the sponsoring advanced economies, a situation sustained by economic processes operating in an interrelated way at the domestic and international levels.

At the domestic level, an important factor in the process was the series of SAP-based devaluation of the local currency, the Cedi (¢). From a stable exchange rate of ¢2.75 to US $1.00 in 1980–1982, the value fell to ¢8.83 to US $1.00 in 1983, the year of adjustment with the first devaluation. With the subsequent frequent devaluations, the *Cedi* equivalent of the US dollar reached ¢1,200.00 in 1995, and by the end of 1997, it had depreciated to ¢2,260.00 to the US dollar. These devaluations rendered Ghana's wood exports on the competitive international market cheaper than before. This is because the wood exporting firms received most of their revenue in local currency. From this revenue in local currency, they were required to repay their SECALs to the commercial banks or the Ministry of Finance and Economic Planning, which guaranteed their respective loans at the latest exchange rate with the US dollar. To maintain their previous levels of real income, as a result of each new round of devaluation imposed, they had to increase their levels of output. Each currency devaluation, as noted by George (1992), therefore, meant that the wood-exporting firms needed to seek ever higher levels of production/ exports in a desperate effort to meet their debt obligations (since the high interest rates and principal amounts were tied to the depreciating values of the cedi).

Despite the government's own ten-year grace period for the repayment of the IDA loan advanced at the beginning of the program, it demanded that the beneficiary firms start repayments within 12 months of receiving the facility, and complete repayments by 1989 (Forestry Commission, 1988, p. 6). There was, thus, tremendous pressure on these firms to increase wood exports to meet the government's deadline, as a matter of survival. In a report by the Forestry Commission on the repayment performance of firms as of February 15, 1988, ten months to the deadline, it was noted that there was "a large number of companies, whose repayment (was) long overdue. Of the 59 borrowers, only 9 (15.3%) had repaid on schedule. 15 (25.4%) had made no repayment at all, whereas although the remaining 35 had been repaying, there is generally large number of Companies whose repayment is long overdue but not yet paid" (Forestry Commission, 1988, p. 6). The survival of the firms at the domestic level was also tied into the dynamics of Ghana's external debt, and the need to boost exports to defray it, because the expansion in processed wood-exports tended to flood the competitive international market, decreasing national revenue from those exports. Like the firms, the government was also forced to ensure still higher levels of production/exports in a desperate effort to meet its external debt obligations. Thus, the government "squeezed" foreign exchange out of the firms and the industry to satisfy international capital by paying off the loans within the shortest possible time, by exerting tremendous pressure on the firms. For the IDA, the pressure ensured that the loans would be paid on schedule. For international capital in general, it ensured that the volume of processed wood exported would increase, and that Ghana's payments on its loans would be assured. In essence, the firms became the government's "work horses" in its quest to bolster its credibility with respect to international capital.

Indeed, the PNDC government's need to satisfy international capital by quickly generating foreign exchange to meet its debt servicing obligations was met before long. Foreign exchange generation by the industry increased immediately after the infusion of the SECALs in 1985 and 1986. In 1987, as much as US $36.218 million was generated by processed wood exports alone. This was enough to defray part of the principal of US $102.099 million on the debt service payments, even after the interest payments had absorbed a US $53.782 million contribution made by the industry to national revenue. In 1990, as much as US $51.723 million was generated in excess of interest payments that could also be used to address a significant part of that year's principal payment of US $139.466 million (World Bank, 1991, p. 75). By June 1990, the government had been able to pay off all remaining recorded external payments arrears, thereby satisfying international capital in terms of arrears, which had been almost US $600 million or about 10% of GDP at the end of 1982 (*Ghana Today* Vol. 1[1] 6.1993, p. 25). The debt service ratio (including IMF and arrears) to exports of goods and services also fell from 58% in 1989 to 40% in 1990. The PNDC government was thus "successful" in getting itself back into the "good books" of international capital, and normal relations were restored with Ghana's creditors, particularly for trade financing in 1990 (World Bank, 1991). The Ghana turf had been re-captured, and even recognized as a "star pupil of adjustment" in international financial circles.

The joint effects of the SAP-based devaluation of the local currency and indebtedness—within the industry and of the economy as a whole—were thus critical in determining export-oriented output in the timber industry. Finally, the removal of the cumbersome pre-SAP exchange control system, as part of the liberalization process, and its replacement with authorized private dealership in foreign exchange (forex) transactions *via* a forex bureaux system, enabled many related international financial transactions to be carried out with ease.

Labor-Related Enabling Factors

The deregulation of the industry, the free market system, and the associated competition forced firms in the industry to minimize costs in order to survive. Consequently workers—skilled and unskilled—were hired or retrenched as was deemed economically expedient (see Owusu, 1994). According to the Ghana Central Bureau of Statistics (1985) by the end of 1982, the firms producing furniture and fixtures employed a total of 1,164 people, including 850 operatives (labor engaged in physical operations) and 66 professionals, 6 of whom were expatriates. By the end of 1984, the sector had lost 249 employees (21.4%), including 197 operatives (23.2%) and 17 professionals (26%), but the number of expatriate professionals had increased from 6 to 9. In the early stages of the program, then, there was a loss of Ghanaian-held positions, either as a result of retrenchment by firms in order to be competitive and thus efficient, or because of workers voluntarily leaving to seek better economic opportunities elsewhere. By the end of the 1985–86 rehabilitation programs, total employment had increased to 1,320, which was 13.4% above the total at the end of 1982. This included 1,093 operatives and 77 professionals and represented increases of 243 (28.6%) and 11 (16.7%) respectively over the figures for the end of 1982. The number of expatriate professionals had also doubled from 6 to 12.

As noted by Owusu (1994), generally, among the furniture and fixtures firms, the large-scale export-oriented companies, employment levels more than doubled, even as they reduced their numbers of expatriate employees. Scanstyle Mim Ltd., the biggest firm and exporter, increased its employees from 220 in 1983 to 540 by the middle of 1991 and reduced its expatriate staff from 5 to 3. The smaller companies like Pideck Ltd. and The Makers Ltd. increased employment of skilled workers to respond to increasing demand. Pideck for example, increased its employment level from 7 to 70. In the case of the government-owned Kumasi Furniture and Joinery Company, the employment level was affected by the government's broad retrenchment policy. For that reason, while it maintained its level of 4 management employees, skilled labor was reduced from 68 to 25 and unskilled labor from 15 to 9. Akuaba Ltd., which produced for both domestic and export markets, for example, maintained its labor size of 120. Its management disclosed that it had a preference for automation but had been constrained by lack of funds. Akuaba, therefore, brought in foreign experts or technicians periodically to keep up its export production. It also disclosed that it had a preference for expatriate production managers, as well as technical experts, because of their "relatively superior knowledge" regarding more complex operations in the export markets and their technical experiences, and their "positive psychological influence" on the Ghanaian employees in terms of engendering productivity. This was because, according to the management, the unskilled workers tended to respond better to foreign bosses, perhaps as a vestige of colonial social relations. Owusu (1994) also observed that the so-called positive influence of expatriate staff on Ghanaian employees, in terms of productivity, was reiterated by the management staff of some other companies, including Atwima Timbers Limited (ATL). An ATL official, for example, disclosed that, for purposes of product quality, the company needed to engage the services of an expatriate to supervise its new line for the production of moldings. This was to ensure that the company's products would sell without problems on the export market at least for a year or two, during which period the foreign expert would be under-studied by Ghanaians. That aside, buyers on the export market, according to the management, tended to have more confidence in companies with such expatriate experts, and such confidence promoted the establishment of trust and goodwill between the firm and the buyers. The use of foreign experts also helped to establish good reputations for companies internationally. Furthermore, the foreign experts or technicians might also have business connections which were beneficial to the company in terms of foreign markets.

Table 5.5 shows the changes in employment levels in the sawmills relative to the rehabilitation programs of 1985/1986. Between the end of 1982 and

Table 5.5. Sawmill Employment Levels Relative to the 1985/1986 Rehabilitation Programs

Period	*End of 1982*	*End of 1983*	*End of 1984*	*End of 1985*	*End of 1986*
Total Employment*	13,278	13,837	13,319	11,865	13,475
Operatives	10,905	11,123	10,764	9,522	10,703
Ghanaian Professionals	395	489	461	360	472
Expatriate Professionals	76	83	77	74	70

Source of Data: Central Bureau of Statistics (1985); Ghana Statistical Service (1989b).
* Includes support personnel such as clerks, typists, and messengers.

the end of 1984, just before the programs began, employment in the sawmills increased, and in fact, increased a little further by the end of the rehabilitation programs in 1986 to support the export drive. Despite this general increasing trend, however there was a net loss of operatives. The number of professionals or skilled labor had increased by the end of 1984 to support the export drive, but fell by the end of 1986. This was mainly because the cost of operations had gone up as a result of the devaluation of the currency, and new equipment had arrived to boost production. There was also a net loss of expatriate professionals because few of them were needed to supplement the operations of the new equipment.

In the specific case of sawmills surveyed by Owusu (1993), the size of management personnel had increased. They were also all Ghanaian. The size of the skilled labor had generally increased in those firms that were doing well like Maxwell Owusu Timbers Ltd., Pan Timbers Ltd., and Timber and Transport Ltd., while the amount of unskilled labor had been reduced in sawmills that were faring badly like Bibiani Logging and Lumber Company. Atwima Timbers Ltd., however, had cut down its skilled labor from 156 to 60 and unskilled labor from 126 to 47, despite its impressive performance to promote efficiency. Management staff, however, increased from 13 to 20 to make management and organization more efficient. In general, the reason given by the firms for the labor cuts was the increasing cost of labor, hence the strategy of increasing the amount of skilled labor, paying them at rates above the market rate as an incentive to retain them, whilst minimizing the level of unskilled labor. Despite the issue of "increasing cost of labor" raised by the firms, production for export was by international standards facilitated by the use of cheap Ghanaian labor. The relative cheapness of Ghanaian labor could be illustrated relative to the price of Coca Cola (Coke). In 1982, before adjustment, the minimum wage was ¢12.00 per day. In comparative terms, a medium-size bottle of Coke was ¢1.80. By 1989, under the adjustment program, the minimum wage had risen to ¢170.00 and a bottle of Coke to ¢120.00. By 1990, the minimum wage had risen to ¢218.00 and Coke to ¢134.70 (World Bank, 1991, pp. 129–30; Statistical Service, June 1991, p. 60). Thus, if a worker bought a bottle of Coke a day, at least half of his/her daily income had been disposed of, leaving less than enough for even a single meal for the day.

On October 1, 1990, the value of the Ghanaian *Cedi* relative to the U.S. dollar was ¢339.00 to US$1.00. The minimum daily wage was thus equivalent to 64.3 cents (UNIDO 1991, p. iii) vis à vis the US$ 34.8 daily minimum in the US at the time. Even though some employers paid a little higher than the minimum wage as an incentive to workers, they seldom paid the equivalent of a dollar, and in fact, the World Bank acknowledged that real wages in most parts of the private sector had actually gone down (World Bank, 1991, p. 15). The employment of expatriate labor was also an important factor in the production for export. While native skilled labor and managerial staff received relatively higher emoluments, their expatriate colleagues made much more. For instance, in Bibiani Logging and Lumber Company's attempt to increase its output and sales, it employed the services of a Polish technical expert, paying him a monthly allowance of ¢200,000.00, as well as a monthly salary of US$5,000.00. The highest-paid Ghanaian management staff-member with an MBA degree from a prominent Canadian university was paid a salary that was just half of the Polish expert's monthly allowance, another indication of how poorly the Ghanaian management personnel were paid. This was more so, since the expatriate personnel were exempted from transfer tax for currencies transferred outside Ghana under the investment code, which guaranteed the employment of approved expatriate personnel, and guaranteed them an exemption from transfer taxes on funds transferred outside the country. Firms, in this regard, were also guaranteed the transferability of such funds as profits and dividends.

Institutional and Other Enabling Factors

To facilitate the easy and rapid export of wood products, the Ghana Timber Marketing Board (TIMBOD), the government agency that was solely responsible for the purchase of timber for export, was replaced by two interrelated bodies, the Timber Exports Development Board (TEDB) and the Forest Products Inspection Bureau (FPIB). TEDB was largely responsible for promoting the sale and exports of Ghana's timber products, while the FPIB was largely responsible for monitoring timber products. The creation of these twin agencies and their organizational structures were recommended by Silviconsult Limited, a consulting firm appointed by the World Bank. This appointment, perhaps, meant that Silviconsult's recommendations would largely be consistent with the economic interests of the World Bank. The TEDB and FPIB, in their operations, subsequently streamlined the hitherto cumbersome timber export processes to make them easier, more efficient, and convenient for producers. For example, the TEDB reduced the number of steps required for exporting processed wood from the required 32 separate steps in 1984, to only 8. Also, it abolished the traditional export license requirement, and made a new and much simpler Exchange Control Form A2 the sole document required for the export of processed-wood. Producers were also encouraged by the regular FPIB monthly and annual reports, highlighting the respective ranks of the various producers in terms of their export performance. These reports generated competition among firms as various managers sought to out-export their counterparts. The replacement of TIMBOD by the twin bodies, according to the FPIB (1991, p. 1), also facilitated "the comparatively high prices achieved by producers from direct sales system between buyer and seller," after the abandonment of the fixed price system used by TIMBOD. The operations of the TEDB and FPIB were enhanced by the activities of other related bodies such as the Ghana Investments Centre, which promoted the very liberal investment code domestically and internationally; and the Ghana Exports Promotion Council (GEPC), which *inter alia*, promoted labor training for wood-processing.

The rehabilitation aspect of the program was also another important enabling factor. This was because it was very heavily biased towards the restoration and development of basic infrastructure related to the export of raw materials as noted by Sawyerr (1990, p. 30). For example, in November 1983, at a conference of Ghana's donors, including the major Western countries like Canada, France, Germany, United Kingdom, and the USA, it was decided that two-thirds of Ghana's foreign exchange expenditure over 1984–1985 was to be spent on rehabilitating the railroad network, roads, and the Tema and Takoradi ports (Government of Ghana, 1984, p. 13). These made the flow of exports out of the country much easier, aided by credits advanced to overcome other physical constraints on access to the forests for the purpose of extracting timber (mainly shortages of equipment, transportation, materials, and spare parts) as shown in table 5.1.

The enabling factors discussed above are largely the "supply" conditions relative to the "demand" conditions that affected production for the export market. On the external front, products from tertiary processing were encouraged because of the very good prices that were then prevalent. In 1991 for example, furniture parts enjoyed a remarkable 55–65% price increase per unit value (FPIB 1991, p. 1) and there was an expansion in markets, which saw an extension beyond the traditional European market to such countries as Japan and Saudi Arabia. The foreign exchange generated as a result of expanded exports served as an incentive for the exporting companies, in addition to the pressures of domestic indebtedness, to augment production for export. This was because such expansion enabled the producers of lumber and other value added products, to retain shares of 20% and 35% of export earnings, respectively, in foreign accounts under the investment code. The strength of the export retention is apparent from the fact that in 1983, the export of lumber yielded US$8.31 million, while furniture and fixtures brought in US$1.04 million, a combined total of US$9.35 million (see TEDB, n.d.; FPIB 1990). In 1990, foreign exchange generated by lumber exports had increased by 903%, whilst furniture and fixtures increased by 619%. Together, they generated US$81.51 million, an increase of 871% over the 1983 figure. The increase in foreign exchange generated was also boosted by an increase in the contribution made by the export of kiln-dried lumber. In 1990 whilst the 1,070 m^3 of kiln-dried lumber exported generated US$572,432.59, in 1991, the volume of 14,934 m^3 exported yielded US$7,161,526.02 (FPIB 1990; 1991). Despite all these enabling factors, Ghana's export-oriented processed wood production could probably have been higher but for a combination of some constraining factors.

Capital-Related Constraints

Generally, returns on investments aimed at adding value to wood products were low with rates of 21% for sawmill improvement, 8% for furniture, and only 4% for wood preservation (Forestry Commission, 1988, 13). Such low returns discouraged further investments aimed at adding value to wood products. For downstream wood processors after saw-milling, this was exacerbated by lack of lumber inputs because the owners of the sawmills preferred to export their lumber output in order to enjoy the tax rebates and foreign exchange retention quotas, based on the volume of export, under the investment code (Owusu, 2001c).

Credits provided under the SECALs such as the IDA/ODA programs directly influenced firms' choice of technology. Even though it made funds available for them for the purchase of plant and equipment, these beneficiary firms' choices were limited by the provisions of the credit agreements such as tied purchases. Purchases by firms had to be products from the donor countries, and in certain cases, the beneficiaries were ignorant of the real nature of equipment and technology they had contracted to buy, or really did not have any options. For example, at Kumasi Furniture and Joinery Company, a joint state-private venture, some idle equipment had to be put on display for sale because it was not suitable for the tropical Ghanaian conditions and therefore useless for production purposes. This was very expensive equipment, which had to be purchased as part of the credit arrangement.

The varied sources of technology and equipment thus made standardization to harmonize production largely impossible, as it had been prior to the ERP. The only limited way standardization was encouraged was through the SECALs. This was because of the few defined sources of equipment and the tied purchases offered (see Hancock 1989), but the amount of such equipment remained small, relative to the older equipment in the industry from the pre-adjustment era, which had been imported by various dealers and merchants. As noted by Barker et al. (1991, p. 33), the lack of standardization presented difficulties for spare parts dealers and mechanics, and also made it difficult to integrate and/or synchronize production equipment. Synchronizing/integrating new standardized equipment from SECAL-related tied-purchases with the old machines presented problems, especially with respect to design specifications, arising from the pre-adjustment dependence on foreign technology and machinery, particularly dependence on used plant machinery and equipment and outmoded technology because such items were cheaper. For the previous foreign owners, it was a convenient way of disposing of old technology for new, at good prices, through sales to firms in developing countries. In fact, special orders had to be placed overseas for custom reproduction of some of the spare parts, which were needed for the pre-used machinery and equipment. This was because the production of such items had long been phased out, and such orders absorbed huge amounts of foreign exchange, even discounting the hold-up time until the orders arrived. The situation was exacerbated by provisions in the investment code, which exempted importers from paying duties on machines or equipment, investment allowances, and depreciation allowances—even on the used machinery and equipment.

Increases in the rate of capacity utilization of equipment among the firms producing furniture and fixtures for export were constrained by poor maintenance and operation of the installed equipment. According to the Ghana Export Promotion Council (GEPC) officer-in-charge of wood processing, as noted by Owusu (2001b) the output of Ghanaian wood processing firms was relatively very low compared with such countries as Indonesia and Malaysia, in spite of the fact that similar types of equipment were employed in each case. Unlike Indonesia and Malaysia, expensive cutters in Ghana, for example, were usually not well maintained and were thus unable to produce the type of finishing required for the export market. It was also noted that generally, factories in Ghana were not batch-production oriented. Products were usually individually processed and manually finished. The machines therefore tended to be usually under-utilized, and processing time wasted. A degree of inertia tended to be reflected in the resistance of producers to batch-production rather than their old piece-by-piece production approach. Retooling for different jobs, the officer noted, also tended to be too time-consuming because effective, integrated organization for the full utilization of machines was lacking, explaining in part the waste of time, and capacity under-utilization. The lack of integration of the installed equipment also led to different depreciation rates for the various machines, leading to frequent breakdowns in the systems and consequently, frequent disruptions in the production process.

Firms producing furniture pieces, furniture parts, and fixtures generally lacked jig-making equipment, a basic requirement for the production of standard furniture components, and this led to the production of large quantities of defective components. Where jig-making equipment was available, it was usually poorly maintained, and thus constrained the quantities and quality of output. In addition, firms lacked access to international standards for components in most cases, a situation that was exacerbated by a shortage of professional local designers. Apart from cases where the overseas buyers provided the subcontracting Ghanaian firms with the specifications for their orders, Ghanaian producers tended to rely on obsolete catalogues from Europe and North America for product designs.

Labor-Related Constraints

By 1990, real wages in Ghana had gone down in most parts of the private sector, including formal wood processing (World Bank 1991, p.15). Low income and rising cost of living tended to generate low morale among Ghanaian skilled workers and management staff. This was exacerbated by income discrimination in firms such as Bibiani Logging and Lumber Company, which favored expatriate personnel vis à vis their Ghanaian counterparts. This demotivated the affected Ghanaian personnel, especially in situations where they felt they had the requisite expertise and could even perform better than the expatriates. Generally, however, a shortage of skilled labor constituted a serious constraint on production and export of processed wood. According to the Ghana Forestry Commission:

> In a basic sawmill, it is often difficult to find skilled saw operators, and engineers with experienced knowledge of kiln drying. Yet, kiln drying is essential for most of the processed wood products. Skilled manpower . . . for the operation and maintenance of moulders, automatic lathes, copy shapers, etc., is severely lacking. Companies engaging in these areas must of need seek to employ key personnel internationally. They tend to have the problem of paying their salaries as well as providing good accommodation and other facilities such personnel will require (Forestry Commission, 1988, p. 13).

The problem of skilled labor shortage continued because not as many people as required were being trained (Owusu, 1995). According to the GEPC officer, whilst the government developed Accra Polytechnic as a model wood-processing school, with plans to improve others, the graduates from the polytechnics were usually unable to meet the skill demands of the industry. This was because inadequate emphasis had been placed on practical, on-machine, training or apprenticeship. Rather, more time tended to be spent on textbooks and classroom lectures. In general, therefore the availability of well-trained machinists was quite limited and thus impacted negatively on production. To address this deficiency, the European Community, the Ghana National Investment Bank, and the GEPC in a technical assistance program to the industry, developed a two-year program to help the firms handle their export orders and execute them competently. This involved a few foreign technical experts working with firms by taking them through the basics of organization and production for export. A saw-doctoring school was also established by the TEDB to address some of the technical problems.

Institutional and Other Constraints

In terms of the external demand for processed wood, there were some strong negative forces or contingent factors, which tended to depress the prices of tropical hardwood. These included anti-wood use campaigning activities of the Plastics Lobby to replace tropical wood with plastics; and the anti-tropical rain forest destruction campaigning activities of environmentalist groups such as Greenpeace. In addition, the European market was generally depressed as a result of recession, and buyers continued to hold large quantities of tropical wood, which moved very slowly. Furthermore, other African exporters had joined in from Central Africa to flood the market with their species, augmenting supply on the world market against low demand, and putting further pressure on Ghana's guiding selling price (FPIB, 1990; 1991).

Indeed, as the foregoing discussion has shown, the production and export of logs and processed wood increased dramatically after Ghana adopted the SAP in April 1983, and the economic measures implemented under the program largely accounted for the change. Effectively, then, adjustment enabled international capital to take back its turf as the PNDC government regained control over its political turf, supported by the Western supporters. This is clear from the overwhelming influence of the SAP-based enabling factors identified in the discussion, relative to the constraining factors. The dramatic increase in Ghana's formal wood processing industry apparently attests to the validity of the proponents' argument regarding expansion in export-oriented production for countries, as they employ the principle of comparative advantage, and use foreign loans to operationalize their export-oriented production process. It also appears to support the view that the production process in developing economies becomes more efficient through competition.

From the perspective of SAP critics, however, the evidence suggests that the changes reflect the PNDC government's commitment to accommodating international capital, as a measure for self-preservation. Consequently, to them, the government had a vested interest in expanded exports, and thus created an enabling environment, provided the necessary support, and generated pressure on firms to achieve them. Processing wood for export was a convenient choice in this case, because it was based on an easily accessible natural resource, and was dominated by lumber, which is essentially a raw material for further industrial processing overseas. This was more so since the annual growth rate of manufacturing as a whole in the Ghanaian economy, mostly consisting in consumer products, fell systematically from 12.9% in 1984 to 1.5% by 1994 (ISSER, 1995, p. 98). In accommodating international capital, the PNDC government had access to more loans, while international capital also reaped profits via increased interest payments on such loans. At the beginning of 1983, the year in which the ERP was adopted, Ghana's total debt was US$1,499.479 million, of which US$1,120.638 million had been disbursed, with a service interest payment of US$38.705 million. By 1987, the disbursed amount had reached US$2.25 billion. Including the undisbursed amount, the debt had reached well over US$3 billion with interest payments having climbed to US$53.782 million (World Bank 1991). In 1991, the total external debt had reached US$3,769.82 million (55.2% of GNP), for which service payments were as much as US$224.00 million and US$119.20 million for principal and interest respectively (ISSER, 1995, pp. 62–64). In 1998, fifteen years into the program, the total external debt had climbed steadily but dramatically to US$5,884.6 million, for which service payments were US$424.9 million and US$138.6 million for principal and interest respectively (IMF, 2000, p. 49).

Even though the two perspectives may be irreconcilable, it is clearly evident that adjustment played a major role in the dramatic increase in the production and export of processed wood. In the next chapter, we shall look at the effects of the ERP on the environment, from the perspective that the SAPs provided a convenient means for African governments and their external sponsors to meet mutual current economic and political interests. These interests are usually met by sacrificing the physical environment through desperate actions such as overexploitation of natural resources so that the immediate convenience of meeting those interests tend to generate huge environmental costs for the adjusting country. This position is tied to two important perspectives on debt and Third World development, the first of which holds that a vital part of the political and economic control mechanisms available to the Western industrialized countries is to keep Third World countries in debt (see Sawyerr, 1990). The second view holds that there is a direct relation between the high foreign debts of developing countries and the rapid rate of deforestation, and that the IMF and the World Bank's export-oriented development model, via adjustment, is largely responsible for the deforestation and environmental damage (see George 1992). In the next chapter, we examine the role of adjustment in deforestation in Ghana and the timber trade.

NOTES

1. An exchange rate of £1 = US $1.50 was used to convert the ODA and ECGD Funds which formed part of the facilities.
2. See the November 16, 1991, issue of the *People's Daily Graphic*, #12746.
3. The GTA was established in 1972 and embraced all in the timber sector—logging, milling, manufacturing and trading. In 1980/81 however, the mill owners decided to break away to form the GTMO headquartered in Kumasi. There have been moves recently by the representative executives to create just one single organization, in the interest of the industry.
4. Quoted from the TEDB brochure, "Ghana: Forests, Wood and People."

Chapter Six

Structural Adjustment, the Timber Trade, and Turf Degradation

Political Convenience and Desperate Deforestation

Ghana's adoption of the free-market based Economic Recovery Program in 1983 at the behest of the World Bank and the International Monetary Fund marked a major policy watershed in its timber trade, with implications, inter alia, for the country's development and the physical environment—the natural turf, namely the geographic or ecological region where the tropical timber grows, and which can be sustained or compromised by human actions or inactions and development. As shown in figure 5.3, and table 5.3, relative to the pre-adjustment years, a dramatic turn-around began to take place in 1983, the year the adjustment program began with a steady expansion in timber exports. The associated economic changes from the program were touted, as we noted earlier, by the sponsors and the international financial community as a big success—a claim that has been widely disputed, particularly by critics of the IMF and the World Bank. Critics such as George (1988, 1992) and Hayter (1989), argue that, on the contrary, SAPs cause, *inter alia,* widespread environmental destruction, while simultaneously emptying the adjusting debtor countries of their natural resources. In addition, the program renders adjusting countries less able to service their debt each year, and precludes investment in economic recovery (George, 1992). Following the Brundtland Report, *Our Common Future* (WCED, 1987), which emphasized the importance of using resources to meet our current needs without compromising the ability of future generations to meet theirs, there has been increasing awareness of global environmental issues. In light of such awareness, a number of non-government organizations like the Environmental Defense Fund and Friends of the Earth have also shown that growth under adjustment, as promoted by the IMF and the World Bank, involves huge environmental costs (see Owusu, 1998).

This chapter argues that Ghana's Economic Recovery Program demonstrates that SAPs provide a convenient means for governments of developing economies and their international capitalist sponsors to meet mutual current economic and political interests. These interests are usually met by sacrificing the physical environment through desperate actions such as overexploitation of natural resources so that the immediate convenience of meeting those interests generates huge environmental costs for the adjusting country. This position is tied to two important perspectives on debt and Third World development, the first of which holds that a vital part of the political and economic control mechanisms available to the Western industrialized countries to maintain control of their turfs in Third World countries such as Ghana is to keep them in debt (see Sawyerr, 1990). The second view holds that there is a direct relation between the high foreign debts of developing countries and the rapid rate of deforestation, and that the IMF and the World Bank's export-oriented development model, via adjustment, is largely responsible for the deforestation and environmental damage (see George, 1992). This chapter therefore contributes to the debate on structural adjustment and environmental consequences in the development process of Third World economies, particularly the relationship between foreign debt and deforestation (e.g., George, 1992; Corbridge, 1993).

THE STRUCTURAL ADJUSTMENT-ENVIRONMENT DEBATE

A major viewpoint in the adjustment-environment debate is that the link between adjustment and environmental destruction is fuzzy. For example, Lensink (1996) pointed out that it is difficult to objectively determine the influence of IMF and World Bank adjustment programs on the environment. His position rested on an earlier study (Opschoor, 1994), which suggested that it is theoretically impossible to predict the direction of environmental effects of adjustment programs. In a similar vein, Reed (1992) concluded in his empirical study focusing on Côte d'Ivoire, Mexico, and Thailand that environmental effects of adjustment programs are really not clear. Such empirical studies of adjustment and the environment—and associated development lessons, particularly for African countries—are conspicuously lacking. In addition to Reed's work, significant work was done by Hogg (1993) focusing on Ghana, the Philippines, and Guyana. Generally, however, writers in this area like Hayter (1989) and Owusu (1994) tended to address the broad environmental implications of adjustment. Unlike such contributions as Reed (1992) and Hogg (1993), this chapter discusses two important aspects of the debate. First, it focuses singularly on Ghana—whose adjustment program was touted as a big success by the sponsors and examines the dynamics of the "export-orientation" of the formal wood processing industry. Secondly, by directly linking adjustment and environmental (rain forest) destruction, in the form of a desperate process of deforestation in the country since the implementation of the ERP, it counters the major viewpoint expressed by writers like Lensink (1996) and Reed (1992).

For our discussion purposes, deforestation refers to any activity in a closed forest which could potentially affect species numbers, or degrade or disturb stand. Any such activity constitutes forest loss, even where a complete conversion or loss of contiguous forest space has not occurred (Osei, 1997). Our main interest here is the activities of commercial loggers, particularly for export purposes, resulting in the removal or destruction of significant areas of Ghana's forest cover and thereby degrading the environment. Using data collected from government and non-government sources in Ghana from beginning October 1991 through June 1997 as well as sources like the World Bank, the World Resources Institute, and the United Nations Food and Agriculture Organization, our discussion focuses on the first decade of Ghana's ERP (1983–1993). We establish the adjustment-deforestation link by a demonstration of how the dynamics of local currency devaluation combine with debt servicing obligations, to exert tremendous pressure on the government and wood exporting firms to systematically increase or maintain high wood export volumes. The burden to produce more tropical wood for export under adjustment constitutes the adjustment-deforestation link. This is based on the premise that the ERP was essentially adopted by the government to accommodate international capital as a convenient measure to meet its own immediate political and economic interests. Sacrificing Ghana's limited rain forest to meet those interests, by way of a sustained and dramatic rapid increase in wood exports that necessarily involved a rapid and dramatic process of deforestation, was the government's recourse in accommodating international capital.

Competing Perspectives: Neo-liberal and Structuralist

In order to examine the adjustment-deforestation link facilitated by the exploitation of timber for trade, there is the need to look at the related debate, since the environmental implications of adjustment constitute a center of controversy between the neo-liberal proponents of the program such as the IMF, the World Bank, the sponsoring countries and agencies, and its primarily structuralist critics, including environmentalist groups such as Friends of the Earth, the Transnational Institute, and Probe International (see Lensink, 1996).

The Neo-Liberal Perspective

The neo-liberal proponents' position rests primarily on the premise that the economic stagnation in Third World countries such as Ghana's pre-adjustment economic malaise, and environmental problems such as forest destruction, were primarily internally generated and sustained (World Bank 1981; Steel

1988). According to the proponents, such problems resulted primarily from market distortions engendered by government interventions in the economy by way of interfering with the "invisible hand" of the market. The solution therefore had to begin from within, using domestically produced resources for which the economically stagnant country has a comparative advantage (as in the production and export of timber), to engage in international trade. Since a stagnant economy lacks the capital to exploit and export the resource(s), international capital would "aid" such countries by providing them with loans like the SECALs. "Growth with debt" should, in such a case, enable borrowers to increase their capacity and to expand with the aid of net foreign savings, for such indebtedness provides the necessary capital used to produce more than a return to cover the cost of borrowing (see Ewusi, 1987, p. 13). In addition, the efficiency-engendering free-play of market forces would, over time, correct the stagnation and the associated environmental problems (see Lensink, 1996, pp. 111–12). After all, from the neo-liberal perspective, efficiency and rational management, based on the market mechanism, are the cornerstones of the increased productivity and profitability expected under the program (see Ould-Mey, 1996, p. 16). For the SAP proponents, therefore, "the invisible hand of the market is the surest path" (see Storper and Walker, 1989, p. 36) to addressing the economic and environmental problems. These arguments are, however, disputed by the critics.

The Structuralist Perspective

The critics' perspective is rooted in the basic structuralist view that problems of Third World economies are largely the consequence of their incorporation into the global market system (Mabogunje, 1981; Sawyerr, 1990). For Ghana the structural problems were initiated during the colonial period, when Britain organized the colony as a tributary economy to expand her own economy. Thus, at the time of independence in 1957, the economy was "warped," characterized by a one-sided external orientation based on the export of raw materials such as timber, and external dependence (see Sawyerr, 1990).

Hamilton (1989) also argued that liberalization measures under adjustment constituted a neo-colonial mechanism for opening developing economies and their markets to western business interests. For example, in the case of Ghana, the technological requirements for the rehabilitation of the wood-processing industry would compel the nation to purchase production, transportation, and telecommunications equipment, as well as their respective spare parts, thereby reinforcing the dependence on imported technology and, to some degree, on foreign technicians and other related personnel. The procurement of such equipment and spare parts would normally be tied to bilateral and multilateral loan schemes as the SECALs for the rehabilitation of the industry. Consequently, the huge foreign debts generated this way would result in deforestation as pointed out by George (1992). This is because, as noted in Lensink (1996, p. 111), the high foreign debts,

> together with the condition related to most adjustment programmes that the indebted countries must meet their debt service, lead to developing countries being forced to increase their production of exports in the short term so that foreign currencies can be earned. Since many developing countries produce relatively cheap products that require the production factor of nature, trade liberalization stimulates overexploitation of the factor of nature.

This "overexploitation of the factor of nature," Ghana's tropical timber in this case, from adjustment's trade liberalization thus results in environmental destruction, as in the unprecedented deforestation of Ghana's limited tropical rain forest after the adoption of the ERP in 1983. According to Sawyerr (1990), the adjustment-based reduction or removal of foreign exchange controls, the lifting of restrictions on imports, and the free repatriation of profits by foreign investors are also instrumental in the process of debt generation and the associated negative environmental implications. In addition, the tied-purchases associated with the loans advanced for the procurement of equipment and spares to exploit the forestry resources (e.g., the obligation to purchase British-made equipment with British loans), and the associated debt obligations, are seen to constitute convenient mechanisms for supporting some corporations and the economies of some donor countries (see Owusu, 1995). In this regard, Hancock in his book, *Lords of Poverty: The Power, Prestige, and Corruption of the International Aid Business* (1989), argued that Britain, for instance, allocated some £850 million per annum to bilateral aid programs. Out of that, some 80% was typically used by the recipient countries to buy British goods and services. He stated that according to the Overseas Development Administration (ODA), multilateral agencies could normally be relied upon to purchase British goods and services with a value equivalent to 120% of Britain's multilateral contributions. For the U.S., he noted that, through the World Bank, 82 cents of every tax dollar contributed are immediately returned to American businesses in the form of purchase orders.

Proponents of adjustment emphasized the efficiency the program engenders in the production process. In fact, this notion of efficiency is not disputed by the critics because the pressure to meet the debt obligations tends to elicit efficient operations at least *in the short run* (Owusu, 1994, 1995). Yet in operations like the export of forest products in Ghana, the positive effects of, or gains from efficiency (e.g., waste reduction or increase in wood recovery rates), were still lost through the negative environmental effects of the increase in timber extraction for export. In the same way, critics agree that prices should properly reflect costs. Yet, the IMF and the World Bank's obsession with "getting prices right" overlooks the domestic environmental costs and the associated negative externalities to the broader global ecosystem. In addition, the devaluation of the local currency associated with adjustment tends to stimulate increased production of raw materials like timber for export because of the depreciated value of such materials on the world market. As George (1992, p. 2) pointed out, however, "(w)ith so many jostling for a share of limited world markets, prices plummet, forcing governments to seek ever-higher levels of export in a desperate attempt to keep their hard currency revenues stable." Infrastructural rehabilitation measures under SAPs were also part of the efficiency promotion process. Critics point out that the restoration or development of basic infrastructure in developing nations such as Ghana normally has a heavy bias towards those related to the exploitation and export of resources. According to Sawyerr (1990, p. 30), such infrastructural rehabilitation measures push the developing countries further into debt with negative implications for the environment, despite the fact that such infrastructure are indispensable to economic recovery. This is because the nature of investments in such infrastructure such as feeder roads into the forests, and the equipment procurement methods adopted in connection therewith, invariably tend to involve the importation of heavy machinery such as timber trucks and earth moving equipment and their spare parts. Contractors and consultants from Western industrialized countries—co-sponsors of the program—are also often required for the construction of such infrastructure including road and telecommunications systems or harbors, which are needed to support the resource export process. Let us now examine how all these issues were pertinent to the Ghanaian economy after the adoption of the adjustment program in 1983. To do this, let us briefly recap the reasons for the program.

Accommodation of International Capital

We have noted that the Ghanaian economy and for that matter the timber industry was in shambles by 1983, and adjustment thus constituted a convenient measure to defuse a domestic political and economic crisis the government itself helped to fashion and to stave off its own imminent collapse as a consequence of its initial failure to "accommodate international capital," by refusing to render Ghana's labor and resources open to exploitation in situ; supplying raw materials and other inputs needed by the Western industrialized economies; and providing markets for their products. For international capital, adjustment would ensure unimpeded access to Ghana's raw materials and markets, as well as uninterrupted movement of capital *into*, and profits and rent *out of*, the economy (Sawyerr, 1990, p. 28). It was thus no surprise that the World Bank (1984, p. xx) pointed out to the PNDC government that "the forestry sector offers the greatest immediate potential for growth and foreign exchange earnings" at the beginning of the program. To simultaneously meet the needs of both international capital and the PNDC government, the forestry sector therefore needed an immediate rehabilitation, and consequently, at the beginning of the ERP, the sector was infused with the second highest sector adjustment loan of US $157 million, out of the initial loans advanced by the World Bank to the PNDC government (Ofei-Nkansah, 1991, p. 229). Further, we noted that the rehabilitation of the forestry sector with adjustment loans constituted a convenient means for the Ghanaian government to satisfy the demands of international capital by keeping the country's hard currency revenues stable via wood and other primary exports. By so doing, the government could also keep itself in power, cushioned by the adjustment loans that the country would receive. For international capital, rehabilitating the forestry sector was a priority because apart from logs, a large proportion of the lumber processed was exported as raw material for international consumption, prior to the deterioration of the sector. Between 1963 and 1969 for example, the sector exported a minimum of 50% of its total output (CBS, 1969), and the years preceding adjustment reflected a similar pattern. Other measures included the removal of existing restrictions on the export of some timber species in log form, imposed in 1979 as conservation and valorization measures, and the replacement of the Ghana Timber Marketing Board with two new agencies, the Timber Export Development Board (for export promotion) and the Forest Products Inspection Bureau (to monitor production) to ensure high export performance (Hutchful, 1996).

As a result of the capital infusion into the sector, there was a dramatic rise in the volume of wood exports, which provided a convenient means for the PNDC government to clear its outstanding external debts, and to meet its new and increasing debt servicing obligations.

Figure 6.1 shows the trends in processed wood production and export from 1973, a decade before adjustment, through 1993, a decade after the adoption of the program. Apart from the significant upward trend after 1983, the proportions of processed wood were significantly greater unlike the pre-adjustment period. This masks the fact that the annual processed wood components in the figure represent only 42% of actual volumes harvested and processed due to the low recovery rates (see Chryssides, 1974; Forestry Commission, 1988). This suggests that overall, very large numbers of trees were harvested, in terms of deforestation, from the forests along with the collateral damage in the forests. As was shown in figure 5.3, a decade into the program (1983 to 1993), increases of 500% (223,914.06 m^3) and 806% (495,613.09 m^3) were

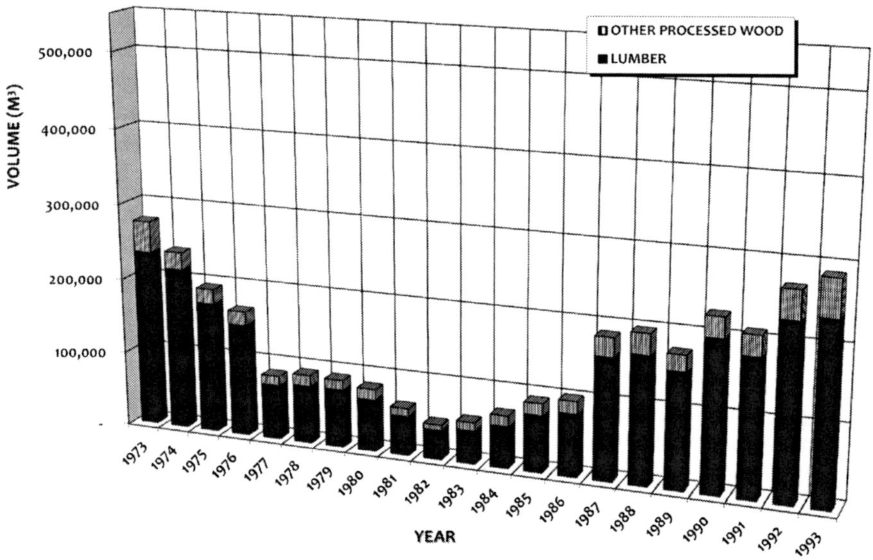

Figure 6.1. Volume of Wood & Wood Products Exported, 1973–1993.
Sources of Data: Timber Exports Development Board, Ghana; Forest Products Inspection Bureau, Ghana; Ministry of Lands and Natural Resources, Ghana

achieved in the volume of lumber and logs exported respectively. Exports initially peaked in 1988 with a total combined wood output of 536,213 m^3 and generated US $98.8 million in foreign exchange (TEDB, 1990, p. 15).

In 1989 there was a significant drop in the export performance of the forestry sector. Relative to the export volume of 536,213 m^3 and revenue of US $107 million of the previous year, those figures fell to 375,769 m^3 (TEDB, 1990, p. 15) and US $81 million (World Bank, 1991, p. 75) respectively. The steady rise in exports from 1990 through 1992 and the sudden jump in 1993 indicate a response to debt pressures resulting from the drop in 1989. In 1991 wood exports amounted to 427,863 m^3 and generated US $114.2 million (FPIB, 1991, p. 1), which was US $5 million less than the US $119.2 million required to meet the interest payment on the external debt for that year. The $5 million shortfall in interest payment required, thus stimulated increased exports of larger volumes of lumber and other wood products, which commanded higher values on the international market, for 1992 and 1993. In order to gain more value from domestic processing, the government was phasing out the export of logs which was expected to be fully in place in 1994. In fact, the log export ban was introduced in 1979 for high value species and extended to all species in 1994 (Bird et al., 2006, p. 2). The actual ban took place in 1996 and even though this official ban was in the pipeline at that time, the volume of logs exported in 1993 even exceeded total export volume of wood for the preceding year, indicating the desperation to meet the external debt obligations. As noted by Birikorang et al. (2001) profit levels within the timber industry rose as a result of the ban, as log prices fell below comparable international log prices whilst exporters continued to sell their products at world prices. As a result of such increased volumes of export, the international market became inundated with wood from other African countries and elsewhere that buyers developed a tendency to stock up the cheap wood which later depressed prices further because of diminished demand relative to supply. With respect to Ghana wood, Barker et al., noted:

> (M)ost of the exports are purchased by only a few buyers in Europe, many of whom resell to other markets in North America and Asia. Although these products are produced in Ghana, they often command prices lower than they would if they were manufactured further or sold to a more diverse market. For example, wood is purchased by companies in Portugal to make doors for sale in the United States . . . lumber is bought by companies in Holland and sold at a profit in the United States (Barker et al., 1991, p. 5).

Also, the recession in Europe at that time led to a very slow movement of the wood stock (FPIB 1991).

The accommodation of international capital is clearly demonstrated by the steady and dramatic increase in the volume of processed and unprocessed wood exported, with the bulk of the increased exports (a minimum of 81% per annum by 1991) going to the donor countries, mostly the then EEC countries. The fact that these raw material exports became so inexpensive that resale often took place on the world market (see Barker et al., 1991, p. 5) underscored the critics' argument that SAPs *inter alia* provide cheap and reliable sources of raw materials for the sponsoring advanced economies. It should also be remembered that lumber and furniture parts that constituted the bulk of the exports are essentially raw materials for firms in the metropolitan or donor countries, and exports from Ghana were fueled primarily by indebtedness, particularly from the SECALs and the frequent devaluation, which cheapened the exports and compelled firms to produce and export more processed wood in order to survive.

DEVALUATION, DEBT PRESSURE, AND EXPANDED TIMBER EXPORTS

In terms of foreign exchange generation, in 1983, the export of wood and wood products yielded US $15.77 million (TEDB, n.d., p. 15). In 1991, five years after the forestry SECALs had been exhausted, wood exports generated US

$114.2 million in foreign exchange (FPIB, 1991, p. 4), an increase of 624% over the 1983 figure. The increase in foreign exchange generation by the sector becomes more significant in relation to Ghana's external public debt and its servicing each year. As a background, by the end of 1982, just before adjustment, the external payments arrears was almost US $600 million or about 10% of GDP (*Ghana Today*, Vol. 1, #1, May, 1993, p. 25). At the beginning of 1983, the year in which the ERP was adopted, the total debt including the un-disbursed amount was just below a billion dollars out of which US $1,120.638 million had been disbursed with a service interest payment of US $38.705 million payment as interest. By 1987, the disbursed amount had reached US $2.25 billion. Including the un-disbursed amount, the debt had reached well over US $3 billion with interest payments having climbed up to US $53.782 million.

Despite the fact that several donors, including Canada, France, Germany, Britain, and the United States, had either forgiven the debts or converted them to grants (World Bank, 1991, p.63), the medium- and long-term outstanding and disbursed external public debt (excluding IMF obligations) at the end of 1990 amounted to US $2.9 billion, or slightly over 50% of GDP, i.e., about the same share as in 1987. Bilateral creditors accounted for about 30% and multilateral creditors for about 65% of World Bank and International Development Association (IDA) loans, and credits amounted to about 40% of the total; financial and suppliers' credits comprised the remainder (World Bank, 1991). In 1991, the total external debt had reached US $3,769.82 million (55.2% of GNP) with service payments as much as US $224.00 million and US $119.20 million for principal and interest respectively (ISSER, 1995, pp. 62–64).

While the forestry sector's possible contributions to the external debt payments increased sufficiently to absorb the interest payments on the loans, the total burden continued to increase at a much greater rate as shown by the rapid increase in service interest after 1985. In terms of foreign exchange generated by the timber industry, in 1984, a year after adjustment, revenue from timber was U.S. $22 million (3.71%) of the total merchandise export (comprising cocoa beans and products, gold, diamonds, manganese, bauxite, timber, and electricity) earnings of U.S. $566 million for Ghana. By 1990, the revenue from timber had increased to U.S. $119 million (13.66%) of the total merchandise export earnings of U.S. $871 million. This was in spite of the world-wide recession (see World Bank, 1991, p.75; 1995, p.75; and FPIB, 1991, p. 4).

The contribution made by the timber industry was relatively low prior to the 1985–1986 rehabilitation programs with US $21 million (3.71%) in 1984 and grew to US $44 million (5.84%) in 1986. Significantly, there was a dramatic change in 1987 reflecting the effects of the capital infusion into the industry for rehabilitation. The share of the sector's contribution jumped for the first time into the double digit percentages, reaching 10.88% from 5.84% in 1986, when the infusion program was completed.

The significance of the foreign exchange generated by the timber industry becomes more apparent, however, when evaluated as a contribution to service interest payments on the outstanding external public debt over the period 1984 through 1991 as indicated by table 6.1. The last column in the table, showing the contribution of the timber industry relative to the annual interest payments, indicates the extent to which the timber sector covers the respective annual interest payments.

The negative signs prefixing the 1984 and 1985 figures indicate that the contributions to interest defrayal were deficient by US $8.238 million and US $2.458 million respectively. On the other hand, the figures prefixed by the plus signs indicate by how much it was in excess. In 1986, for example, foreign exchange generated by the timber sector was enough to fully absorb the interest payment of US $42.624 million with an excess of US $1.376 million, and reduce that year's principal repayment requirement of US $79.274 million.

For the PNDC government, therefore, the need to satisfy international capital by quickly generating foreign exchange to meet its debt servicing obligations was met before long. Foreign exchange generation by the sector increased immediately after the infusion of the SECALs in 1985 and 1986. In 1987, as much as US $36.218 million was generated by processed wood exports that could defray part of the principal of US $102.099 million on the debt service payments, even after the interest payments had absorbed a US $53.782 million contribution made by the industry. In 1990 there was as much as US $51.723 million in excess of interest payments that could be used to address a significant part of that year's principal payment of US $139.466 million (World Bank, 1991, p. 75). It is therefore not surprising that by June, 1990, the PNDC government had been able to pay off all remaining recorded

Table 6.1. Foreign Exchange Generated by the Timber Industry as Contributions to Interest Payments on Ghana's External Public Debt (Outstanding): 1984–1991 (US $ millions)

	Debt Service Payment			
End Year	*Principal*	*Interest*	*Revenue From Timber Exports*	*Contribution Relative to Interest*
1984	64.646	29.238	21.0	–8.238
1985	67.243	30.458	28.0	–2.458
1986	79.274	42.624	44.0	+1.376
1987	102.099	53.782	90.0	+36.218
1988	138.388	69.166	107.0	+37.834
1989	147.444	59.499	81.0	+21.501
1990	139.466	67.277	119.0	+51.723
1991	224.000	119.200	114.2	–5.000

Sources: World Bank (1991, 75, 99), FPIB (1991, 4), and ISSER (1995, 63).

external payments arrears, thereby satisfying international capital in terms of arrears. The debt service ratio (including IMF and arrears) to exports of goods and services also fell from 58% in 1989 to 40% in 1990. The PNDC government had thus been "successful" in getting back into the "good books" of international capital, and normal relations were restored with Ghana's creditors, particularly for trade financing in 1990 (World Bank, 1991).

The above "successes" in terms of the dramatic increases in exports, in foreign exchange generation, and the recognition of Ghana as a "star pupil of adjustment" in international financial circles are all tied to the pressures unleashed by a combination of indebtedness and the devaluation of the local currency—the *cedi* (¢). In other words, the Ghanaian government was under pressure to meet its debt obligations in order to win the "seal of approval" of international capital. At the same time, it had to do so under the pressure of a series of adjustment mandated devaluation of the *cedi*, whereby prices plummet, forcing the government to seek ever higher levels of exports in a desperate attempt to keep the hard currency revenues stable (George, 1992). From a stable rate of ¢2.75 to US $ 1.00 in 1980–1982, it jumped to ¢8.83 in 1983, the year of adjustment with the first devaluation, reaching ¢1,200.00 in 1995 and had risen to ¢2,260.00 by the end of 1997. Each devaluation of the *cedi* meant that the timber firms needed to seek ever higher levels of production/exports in a desperate effort to meet their debt obligations (since the high interest rates and principal amounts were tied to the depreciating values of the *cedi*). Also, the pressure on the firms was aided by the deregulation of the industry and the free and open market system forced the firms, as a measure for efficient operations, to minimize costs in order to survive.

It is, therefore, under this condition of a vicious circle of debt pressure, illustrated in figure 6.2, that Ghana's rain forest began to be "mined" for its tropical wood in a desperate, unprecedented fashion, a process assisted by a lack of a coherent policy of forest management (see Hutchful, 1996, pp. 178–79). On the domestic scene, the pressure to repay the SECALs by the beneficiary firms did play a significant role in the generation of foreign exchange for Ghana over the period as shown in figure 6.2. Generally, the performance of most of the beneficiary companies, with respect to their loan repayments, was very poor. Despite the government's ten year grace period for the repayment of the IDA loan advanced at the beginning of the program, the beneficiary companies were asked by the government to start their repayments within 12 months of receiving the facility. They were also required to complete repayments by the end of 1988. There was thus a tremendous pressure on the beneficiary firms to increase wood exports in order to meet the deadline set by the government. For the IDA, it ensured that the loans would be paid by the end of their ten year moratorium and for international capital

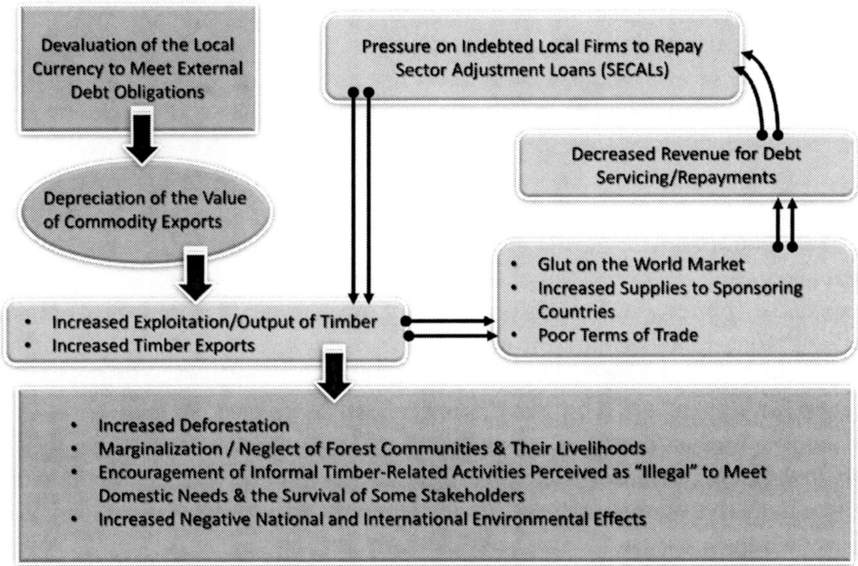

Figure 6.2. Devaluation, Debt Pressure, and Deforestation
Source: Adapted from Owusu (1998b, 2006).

in general, it ensured that the volume of processed wood exported would increase and that Ghana's payments on the debts would be assured. It was also an opportunity for the PNDC government to recycle the paid loans before the scheduled payment time, in addition to its credibility with respect to international capital. In essence, the beneficiary firms became the "mules" for the government, and again as figure 6.2 shows, those firms constituted an important part of the vicious debt pressure circle. There was, thus, a paradoxical situation with regard to foreign exchange generation by the wood processing industry, under the ERP: Whilst the industry as a whole performed very well in generating foreign exchange for the country, the individual firms found themselves heavily indebted and found it rather difficult to keep up with their debt service payments. In fact, there was generally a large number of companies whose repayments were long overdue (Forestry Commission, 1988).

Since the rapid increase in the generation of foreign exchange by the wood-processing firms under the program was to a great extent driven by the pressure to increase output and exports in order to service the loans they had taken, one could suggest that the government, really, was "squeezing" foreign exchange out of the firms and the industry, as a way of satisfying international capital by paying off the loans within the shortest possible time. As noted earlier, the government took the SECALs at interest rates between 8%

and 10%, with a ten year moratorium from the donor countries. However the government turned the SECALs into commercial loans for the firms with the concurrence of the donors, with interest rates of 31% to 35% and a four year repayment period. The SECALs were advanced without working capital and therefore created heavy pressure on the firms. Supplementary overdrafts were also hard to secure from the local financial institutions because of the financial structure reforms under the ERP. Where overdrafts were available, they attracted an interest rate of 27%. These were in addition to the fact that the machinery and equipment purchased by the firms for rehabilitation were quite expensive (Forestry Commission, 1988). For example, a kiln drying system of 300 m^3 capacity imported from Europe, cost over US $300,000.00 (Barker et al., 1991, p. 31) relative to an average of US $1.75 million in SECALs advanced to nine sawmills for plant improvement (Forestry Commission, 1988).

As part of the creation of an enabling environment for investments by local and foreign capital under the program, the wood processing firms were given incentive packages under the investment code (PNDC Law 116, 1985) to encourage exports. This also added to the problem since they discouraged sale of lumber on the domestic market. The sawmill owners and managers preferred to enjoy tax rebates based on volumes exported. Additionally, they could retain at least 35% of the export earnings in an external account (for safe keeping), as well as an extra portion of the earnings, ostensibly for the purchase of machinery, equipment, and spare parts. The dramatic increase in volumes of wood exported since 1983 and the quick income generated by the forestry sector certainly met the needs of the government and the external sponsors in a convenient way. However, generating such quick income by falling on the forestry sector, as a matter of convenience, necessarily entailed rain forest deforestation; and the tremendous pressure on the government and firms to meet their debt service obligations necessarily translated into desperate deforestation.

ENVIRONMENTAL COST OF THE EXPORT EXPANSION: UNPRECEDENTED TURF DEFORESTATION

The market-driven hunt and extract system of logging to meet export demands, as part of Ghana's accommodation of international capital, strengthened as a major component of deforestation in Ghana. The process necessarily involved loss of the desired trees, as well as loss of other flora and fauna through collateral destruction and degradation. The affected areas generally tended to be subsequently utilized for agricultural purposes and other human activities. It is in this vein that the Trade and Environment Database (TED) of the American University (TED case Number 193) suggested that Ghana's timber trade had been the main source of deforestation in the country since the beginning of Ghana's adjustment, as evidenced by the dramatic increases in exports. As noted by Osei (1997), commercial logging encourages the construction of roads into previously inaccessible forest areas and such roads tend to be widely used by "slash and burn" farmers to access new areas of farmland, thus expediting the rate of forest loss. This is particularly so in the case of farmers who can no longer use the rich lands that have been converted for production of export crops stimulated by adjustment (George 1992). Ghana's case was similar to neighboring Côte d'Ivoire, where Chege (1994) has shown that for every 10 km^2 of forest area logged, 10 km of road was constructed. In fact based on FAO assessments, the TED suggested that the rate of deforestation in Ghana increased by 50% over the ten years of adjustment, thereby driving some of the traditional export species to the point of extinction (also, see Hutchful, 1996, p. 179).

Fuelwood consumption is often identified as a leading cause of deforestation in Africa (see WRI 1994, p. 305), but it is really not a significant factor in Ghana, where the annual consumption is some 650,000 tons (TEDB, undated). This is because, unlike countries such as Burkina Faso and Chad of the drier region, under the normal, traditional wood collection system in Ghana, deadwood, generated via attrition in areas of various stages of forest succession, are harvested for fuel rather than live trees (Osei, 1993). Fuelwood is also generated from abandoned unprocessed trunks, after loggers (and sawmillers) have sawn off their contract sizes. The annual charcoal consumption of 120,000 tons in Ghana (TEDB, undated) is also generated the same way, thus making the commercial loggers the main source of the deforestation. In fact the fuelwood-processors like the slash and burn farmers also take advantage of logging roads to penetrate the forest (Hutchful, 1996). It is not surprising that the percentage increase in the extent of Ghana's cropland, as a reflection of a desperate deforestation, was 23% between 1983 and 1993 as compared with 16.7% and 6.1% of Côte d'Ivoire and Nigeria respectively (see Allen, 1997, p. 90).

To put the extent of Ghana's deforestation in perspective, it should be borne in mind that the approximate original extent of Ghana's closed tropical moist forest (i.e., all areas that were naturally forested before intensive exploitation set in) was 145,000 km^2 as shown in table 6.2.

Ghana was actually one of the countries with the highest extent of closed moist forest in West Africa before the advent of intensive logging beginning in the early 1890s, for export to Western Europe. This was after Britain gained control of much of southern Ghana as part of the colonial process. As noted earlier, this timber trade coincided with the beginning of cocoa cultivation for export, which also involved deforestation (Dickson, 1969).

Table 6.2. Original versus Remaining Extent of Closed Canopy Moist Forest in West Africa

Region/Country	Approx. Original Extent (km^2)	Remaining Extent (km^2)[a]	Percent Remaining[a]
West Africa[b]	1,251,600	182,180	14.6
Benin	16,800	470	2.8
Côte d'Ivoire	229,400	44,580	19.4
Ghana	145,000	17,180	11.8
Guinea	185,800	20,500	11.0
Liberia	36,100	20,000	20.8
Nigeria	421,000	59,500	14.1
Sierra Leone	71,700	7,400	10.3
Togo	18,000	3,040	16.9

Sources: IUCN (1992); Osei (1997).
Notes: [a]From FAO (1988). [b]Including Senegal and Gambia.

By 1988, the United Nations Food and Agriculture Organization (FAO) noted that the remaining extent of the closed tropical moist forest was 17,180 km^2 or 11.8% of the original extent. In a booklet, *Ghana: Forests, Wood and People*, published in the early 1990s, the government's official timber export promotion agency, the Timber Export Development Board (TEDB), noted that the existing forest extent comprised 17,000 km^2 of forest reserves and 3,700 km^2 of unreserved forest. Based on those official government figures, the Forestry Research Institute of Ghana noted in 1992, that "Ghana may not have any forest by the year 2030" if the rate of deforestation under the ERP from wood exports continued (GNA, 1992, p. 1).

As shown in table 6.3, even though the World Resources Institute (WRI, 1994) and the World Bank (1997) indicators show higher figures, with respect to the extent of the forest, they still confirm that the extent of the forest had diminished dramatically since the 1980s and logging has played no small part. The table also shows that Ghana had the second highest annual rate of natural forest deforestation in West Africa, and registered the highest loss of 13,000 km^2 from 1980 through 1990. This must be seen in the light that rehabilitation of the forestry sector (using logging equipment purchased with SECALs) began earnestly in 1985, suggesting an extraordinary intensive logging—"unprecedented deforestation"—from 1985 onwards. As shown in figure 6.1, 1985 marked the beginning of the dramatic rise in wood exports, in contrast with the two years preceding adjustment in 1983. Also, logging 110 km^2 annually from 1981 through 1990, as table 6.3 shows, is significant in terms of deforestation in a fast dwindling forest area, especially in the absence of a coherent forest management system and policy. In 1988, the World Bank itself admitted that "in spite of the importance of Ghana's for-

Table 6.3. Changes in the Extent of Ghana's Natural Forest, 1980–1990

	Extent of Natural Forest ('000 km²)			*Annual (1981–1990)*	
Region/ Country	*1980*	*1990*	*Net Change 1980–1990*	*Deforestation Rate (%)*	*Extent of Logging (km²)*
West Africa	615	556	–59	1.0	3,120
Benin	56	49	–07	1.2	0
Côte d'Ivoire	121	109	–12	1.0	850
Ghana	109	96	–13	1.3	110
Guinea	76	67	–09	1.1	90
Liberia	49	46	–03	0.5	790
Nigeria	168	156	–12	0.7	1,270
Sierra Leone	20	19	–01	0.6	10
Togo	16	14	-02	1.4	10

Source: World Resources Institute (1994).

est resources . . . no up to date and comprehensive policy" existed for the sector (World Bank 1988, quoted in Hutchful 1996, p.179). Again, in terms of the sustainability of that rate of logging, a World Bank staff appraisal of the sector noted that the actual cuts were far above the allowable cut "by approximately 1.6 to 2.5 times the current allowable cut"—1.8 to 2.7 million m^3 as opposed to 1.1 million m^3 (World Bank, 1988, quoted in Hutchful, 1996, p. 179). Despite the long gestation periods of some species (200 to 500 years in some cases), reforestation, natural or tree replanting by the logging companies, is insignificant (Owusu, 1994). The haphazard nature of the hunt and extract logging process does not open the canopy sufficiently for many of the important tree species to regenerate naturally (Barker et al., 1991, p. 26).

A 1988 Report by the Forestry Commission noted that ". . . heavy logging has generated public concern about the rate of timber exploitation" (Forestry Commission, 1988, p. 12). In 1992, the proprietor of The Makers, one of the wood processing firms remarked to me that "the government has been exploiting and exporting timber as if we are the last generation of Ghanaians." The author returned to Ghana in June 1997, and one of his first observations, as compared with 1992, was the relatively smaller girths of timber logs that were being carted on trucks from the forests to the mills and to the harbors for export. In fact, the author hardly notices logs of large girth being transported to various sawmills during his annual visits since 2000. This shift to the logging and subsequent processing and export of much leaner timber trees underscores the desperate nature of the deforestation process as the government struggled to meet its debt obligations. According to the World Bank (1997, p.246), Ghana's total external debt rose from US $1,398 million in 1980 to US $5,874 million in 1995. Debt service as a percentage of exports of goods and services rose from 13.1% in 1980 to 23.1% in 1995, and multilateral external debt as a percentage of external debt, also rose from 19.9% in 1980 to 50.8% in 1995. These figures mean that there was more pressure on the government to meet its debt obligation, and the sacrifice of Ghana's forest as a matter of convenience may well continue until the "special goose"—the forest—is unable to deliver the "golden eggs" any longer as a result of depletion or stark environmental damage.

To prolong the convenience and to stave off pressure from environmentalist groups on the government and international capital, as was the case in the international ivory trade, to save the forests, Ghana initiated the Forest Resource Management Project (FRMP), a conservation plan in 1988. For the government, turf-wise, the project reflected its apparent concern for protecting the forest resource base and coping with forest depletion and environmental degradation. The support for the project came in 1989 from the Danish International Development Agency (DANIDA), *inter alia*, and from none other than the ERP-sponsoring international donors, namely the International Development Agency of the World Bank (IDA), and the Overseas Development Administration of the United Kingdom (ODA), who necessarily had to maintain their turf, because of their vested interest in the natural resource. In that regard, the FRMP focused mainly on institutional strengthening of the Forestry sector agencies including infrastructure development, training, development policy planning, monitoring, and evaluation ability (MLF, 1996b).

Until 1994, there were really no clearly defined and detailed forest policies that specified the goals, objectives, and strategies for forest development and the future direction of the timber industry. This led to the emergence of new turf dynamics as the government, concessionaires, farmers, and illegal loggers staked their respective turfs by their actions, reactions, or inactions. In that context, illegal forest harvesting was encouraged because of low fines and lack of legal sanctions (MLF, 1996b), and as noted by Boateng (1991), illegal cutting of timber and forest encroachment for farming purposes intensified the process of forest degradation. This was exacerbated by the alienation of forest communities from policy formulation, despite the expectation that these communities would help to protect the forests (MLF, 1996b). Concessionaires also contributed to the chaotic situation by selectively felling only preferred commercial timber species (Forestry Commission, 1994).

GOVERNMENT POLICIES FOR RECOVERY: THE 1994 FOREST AND WILDLIFE POLICY

The FERN Report (2006) succinctly describes the factors that pushed the NDC government to address the environmental problems that had been unleashed as follows:

> By the early 1990s, several factors had combined to compel a fundamental policy review. There was global (particularly donor country) concern about the environment, especially climate change and the importance of tropical forests in combating this; there was heightened national awareness of deforestation and its consequences following the major drought and bushfire crisis of the mid-1980s; and finally, the public became aware of the central role the timber industry played in deforestation. Indeed, from the late 1980s through the early 1990s, Ghana was at the center of a series of spectacular exposés of the corruption and rapaciousness of the international timber industry. All parties therefore had an interest in restoring order and credibility to the sector. Obviously, the reformers either miscalculated the state's real agenda or underestimated the strength of industry-state symbiosis (FERN, 2006, p. 21).

In specific terms, in 1994, the government finally realized what was actually happening to its turf and the policies in place to address the related issues. It became apparent to the NDC government that most of the provisions in the old forestry policy of 1948 could not adequately address the emerging environmental and the associated economic and social realities any longer, and therefore adopted a new National Forest and Wildlife Policy geared at providing "an additional basis to develop a national forest estate and a timber industry that provides the full range of benefits required by society in a manner that is ecologically sustainable and that conserves (the) environmental and cultural heritage." The new policy also sought to promote "public participation in the share of benefits and responsibilities in forest management" and encourage "integrated coordinated research in forest-related issues" (see Forestry Commission, 1994). Key aspects of the policy were to:

1. manage and enhance Ghana's permanent estate of forest and wildlife resources;
2. promote viable and efficient forest-based industries, particularly in secondary and tertiary processing;
3. promote public awareness and involvement of rural people in forestry and wildlife conservation;
4. promote research-based and technology-led forestry and wildlife management, utilization, and development;
5. develop effective capability at national, regional, and district levels for sustainable management of forest and wildlife resources.

Important aspects of the new realities, as we noted above, were widespread illegal harvesting and disregard for established harvesting (Donkor, 2003). To make the new National Forest and Wildlife Policy operational, a medium-term Forest Development Master Plan (FDMP) was developed for the period 1996 to 2020. This master plan was couched in the National Medium Term Development Plan—Ghana Vision 2020, geared at raising the country's living standards to the level of a middle-income country by the year 2020 (MLF, 1996b). This master plan was designed to be driven by the private sector, mainly through the reforestation plans required by a new concessions system. It targeted, *inter alia,* the development of sustainable forest and wildlife resources, the modernization of the timber industry, and the conservation of biodiversity and the environment. It was also designed to address other specific commitments regarding:

1. The introduction of a forest management certification system;
2. The merging of the Timber Export Development Board and the Forest Products Inspection Bureau into the Timber Industry Development Board;
3. The implementation of 1988 Timber Resources Management Act provisions and regulations (Act No. 547/88) designed to introduce social responsibility, local participation, and replanting into timber concession management.

Regarding the third item, the government then issued two legislative instruments, the Timber Resources Management Act 547 in 1997 and the Timber Resources Management Regulations in 1998 to guide the allocation of timber rights within and outside reserves. The Timber Resources Management Act was geared at consolidating and replacing all existing forestry legislations. It proposed clear identification of land and forest-holding communities as the primary clients of a proposed Forest Service, which is expected to pursue sustainable forest management. The Act was also used to introduce substantial innovations concerning forestry operations concessionaries, particularly, the Timber Utilization Contract (TUC).

The TUC is a contract between the government and the operators comprising only registered companies, subjected to the presentation of: reforestation and restoration plans, inventory of efficient equipment, listing of qualified staff as well as a social responsibilities agreement with landowners. Timber Utilization owners are required to submit reforestation plans for the TUC area and re-planting at the rate of 1 km^2 for each 10 km^2 exploited. To qualify for a contract, a Timber Rights Evaluation Committee (TREC) has to certify that the concessionaire is capable of reforesting 10% of the area granted as a concession. As an incentive, the planted timber becomes the property of the concessionaire. The committee however ensures that the concessionaire has efficient processing facilities to reduce waste, as well as professional foresters on staff, and plans to address likely negative environmental effects, and agreements with land owners to meet related social responsibilities (see Donkor, 2003). Also, as a penalty for excessive use of the popular species to promote the use of less used species, the legislative instrument on Act 547 raised stumpage fees and royalties. The fees were tied to the star rating of the species. The highly used species (scarlet star-rated) attracted 20% FOB of air dry lumber prices; normally used species (red star-rated) attracted 10%; and lesser used species (pink star-rated) attracted 5%. The increased cost elicited improved efficiency in the utilization of the timber (Donkor, 2003). In 1998, The Ministry of Lands and Forestry also started the development of a Forest Management Certification System for the country and a comprehensive computerized system for log tracking with assistance from EU and the Government of The Netherlands (Bird et al., 2006).

It must be noted that in the main, the policy developments in the industry are linked to the Year 2000 Objective of the International Tropical Timber Organization (ITTO) of which Ghana is a member (ITTO, n.d.). The Year 2000 Objective was a declaration in 1990 by the ITTO member countries that "total exports of tropical timber products should come from sustainably managed forests by the year 2000." This was then reaffirmed in 1991 with the definition as a "strategy by which through international collaboration and national policies and programs, ITTO members will progress towards achieving sustainable management of forests and trade in tropical timber from sustainably managed resources by the year 2000." The role of the ITTO is significant in the sense that it represents another player in the tropical timber trade with a turf composed of the member countries. This is more so since it was originally conceived as a commodity organization on a global scale to promote the trade in tropical timber but has in practice become an agency for sustainable development through trade, conservation, and best practice forest management. Such agency may conflict with the agency of international capital in its search for profits as the two turfs overlap, especially as the ITTO's Libreville Action Plan of 1998, inter alia, calls for the improvement in the use of the resource to give greatest possible social benefit, and limiting timber harvesting to sustainable levels, both of which may not be efficient from a neo-liberal perspective.

The foregoing discussion established a direct link between adjustment and environmental destruction, as illustrated by the vicious circle of debt pressure in figure 6.2, and the associated desperate deforestation process has been a convenient means for the Ghanaian government and international capital to meet some current mutual needs. For the government of Ghana, the adjustment program and the contributions of the forestry sector have been a convenient means of accessing international loans by actively servicing the country's debts with foreign exchange generated through the export drive. In 1986 the share of official debt (i.e., government and government guaranteed debt) in total long-term debt was 85%. In 1992, that percentage had climbed to 93% (World Bank, 1993) indicating increasing indebtedness as result of easier access to international loans. In accommodating international capital via the program, the PNDC government was able to stay in power by defusing the political crisis it faced prior to adjustment. It was even able to transform itself from a military government into a "democratically elected" one—the National Democratic Congress (NDC). It enjoyed "political stability" with support from the West, and become a protégé of the sponsors, making Ghana a "star pupil of adjustment" in the eyes of the World Bank and the IMF.

For the external sponsors or international capital, the adjustment program and the contributions of the forestry sector thus became a convenient means of generating cheap raw materials—tropical wood and wood products—in abundance, for their home industries. The loans advanced to Ghana by the sponsors, and the interests on them were serviced consistently. The sponsors were also able to control decision-making in the Ghanaian economy. Furthermore, the wood processing sector served as a convenient market for products like machinery and equipment from some of the donor countries, and the administration of the SECALs was a convenient way of sustaining, and in fact reinforcing, the market. This was particularly successful with the system of tied-purchases and credit-line system as noted by Hancock (1989). In addition to selling products like sawmilling and other wood processing equipment to firms in the sector, some of which were inappropriate for conditions in Ghana, spare parts had to be subsequently obtained from the same sources for their maintenance. In certain cases, technicians from the seller-companies had to be brought in at the expense of the buyers to either install or repair the machinery or equipment. In quite a number of cases, the equipment purchased broke down quite frequently. The tied-purchase and credit-line systems thus propagated what has been described as an international "junk-yard syndrome," whereby inappropriate and substandard machinery and equipment or technology are dumped in developing countries for profit (Owusu, 1994). The sector thus, became a market of convenience and dumping grounds for products from the donor countries as well.

Before adjustment, the forestry sector was in a state of decline with deteriorated equipment and poor production levels, a situation which dramatically changed with capital infusion, sectoral reorganization, and rehabilitation under adjustment. While increased production and exports since 1983 confirm improved efficiency, as asserted by supporters of adjustment, the important question is who benefited from the program's "success" and its underlying "efficiency"? Were the negative environmental consequences—domestic and the externalities—of the desperate deforestation factored into the "success" of Ghana's program? Clearly, the real (short-term) beneficiaries of the program were the sponsors/international capital, reinforcing the view that "multilateral financial institutions, as agents of 'multinational imperialism', work to ensure that the economies of the Third World are conditioned to 'accommodate' international capital, i.e., serve the basic interests of Western industrialized economies" (Sawyerr, 1990, p. 28).

As asserted by adjustment supporters, the right prices may have been restored via the invisible hand of the market; market forces may have fostered efficiency in the forestry sector; and the application of the theory of comparative advantage may have helped to put more of Ghana's wood on the international market. Certainly, however, the overall negative environmental consequences of the desperate deforestation process were not factored into the "success." For critics like George (1992) and environmental groups like Friends of the Earth (1993), and as pointed out in the TED Case Number 193, the "success" was achieved at the cost of a squandered environmental capital and economic future—with serious national and international implications. To meet the immediate needs of international capital and the Ghanaian government, Ghana's limited tropical rain forest was sacrificed, but this mutual "immediate convenience" for the PNDC and NDC governments and international capital certainly translated into "desperate deforestation," which environmentally, was seriously jeopardizing our common future.

In evaluating the 1994 Forest and Wildlife Policy, the FERN Report (2006) pointed out that it did not mark a sea change in Ghana's forest governance; indeed, it represented only a temporary set-back to resource plunder, and that it really was not a clear guide to the corrective action that the authorities promised. Rather than choosing between fundamental paradigms for forestry, it gave the impression that all agendas can coexist, by reading rather like a menu of competing interest groups' agendas, and failed to engage sector-specific problems. To the extent that forest regulators and politicians paid any attention to it at all, they clearly read it as license to continue business as usual. In other words, it preserved the status quo by throwing competing agendas in the mix (FERN, 2006, 20–21).

In order to regain the turf sacrificed by the PNDC and the successor NDC governments for convenience, the new government of the National Patriotic Party (NPP), which came to power in 2000, was forced by the fall-out from the neo-liberal adjustment strategy to adopt a more comprehensive and relatively holistic strategy by developing and implementing new policies rooted in the United Nations sanctioned ITTO's Year 2000 Objective, a declaration made by ITTO member countries in 1990 (with some changes in 1991, 1994, and 1998) to source all exports of tropical timber products from sustainably managed forests by the year 2000. In January 1996, the Ministry of Lands and Forestry directed that the annual allowable cut for timber be reduced from 1.2 million m^3 to 1.0 m^3, the rationale of which was to curb the excessive exploitation of the natural forest, which had reached an alarming level by that time (Donkor, 2003). On the international arena, Ghana was listed in

2000 as one of the top six countries—"The Super Six"—which appeared "to be managing some of their forests sustainably at the forest management unit level to achieve the Year 2000 Objective" although it has not as yet achieved "full implementation in the forest" (ITTO, 2000; Poore and Chiew, 2000). In September 2001, the new NPP government launched the National Forestry Development Program with an ambitious annual planting target of 20,000 hectares. This program relies on private sector involvement, particularly through agro-forestry *taungya* practices by local farmers on areas of degraded forest reserves—a system of forest plantation in which small-scale farmers are allowed to cultivate crops between the seedlings of a forest plantation for the first few years after planting (Amanor, 1999).

Within the national arena, however, turf conflicts characterized the industry both during the adjustment era and under the NPP government. As we shall see in the next chapter, the neo-liberal adjustment program undermined spatial integration in Ghana by destroying the traditional link between the formal and informal sector of the industry, whilst the Ghana Timber Millers Organization, for example, argues that the timber industry, an important component of the economy is on the brink of collapse as a result of the NPP government's extensive interventions and levies (GTMO, 2000; Donkor and Vlosky, 2003).

Chapter Seven

Turf Dynamics and Conflicts, the Timber Trade, and Spatial Integration

With a primary focus in the environment, we continued our discussion of structural adjustment and its implications in terms of its impacts on the Ghanaian economy relative to turfs and the tropical timber trade in chapter 6. We argued that Ghana's Economic Recovery Program demonstrates that SAPs provided a convenient means for African governments and their international capitalist sponsors to meet mutual current economic and political interests. We showed that these interests tend to be met by sacrificing the physical environment through desperate actions such as overexploitation of natural resources—tropical timber in this case—so that the immediate convenience of meeting those interests tend to generate huge environmental costs for the adjusting country. To regain control of, and sustain the hold on their respective turfs, both international capital and the new national government cooperated in certain respects, in the case of Ghana to formulate new policies to ensure continued access to the timber and other natural resources.

In this chapter, we focus on the new trends and associated patterns of the timber trade in terms of sourcing, volume, content and direction, and the associated turfs and dynamics that emerged as a result of the program. We show that generally, the program engendered a dramatic up-turn and expansion in Ghana's timber exports, especially lumber, to the traditionally dominant European market, with new markets, including the Middle East, Asia/Far East, North America, and some West African countries also steadily becoming important. We discuss the factors or determinants that engendered the upswing in production and exports and the associated new trends. These new trends however are shown to have had largely negative effects not only on the environment as we saw in chapter 6, but also as will be shown, on the traditional interdependent relationships between the formal and informal sectors of the wood industry and on the very livelihood of the forest communities and forest management. At the domestic level, it is shown that Ghana's quest to sustain its wood exports has constricted the traditional forward linkage between sawmills and downstream processors and created an acute lumber shortage on the domestic market, thus generating what is largely perceived to be illicit activities by chainsaw operators to meet the domestic demand and further marginalizing the forest communities. At the international level, it is shown that Ghanaian wood processing firms now operate as subcontractors of convenience for firms overseas via an expanded forward linkage, and in the process exporting potential Ghanaian wood-processing jobs overseas.

The discussion in this chapter is couched in the view that a healthy symbiotic relationship between the formal and informal wood sectors in the economy as well as the forest communities is critical for the sustainability of the timber industry, as a whole, and the associated domestic and international trade. Fostering such a relationship requires a new, pragmatic, and sympathetic look, by African governments and policy makers, at the true nature and dynamics of the informal sector and the forest communities. This way, policies formulated and implemented would be based on partnership, a necessary ingredient for the sustainability of the forest sector as a whole.

STRUCTURAL ADJUSTMENT, THE FORESTRY SECTOR, AND THE CHANGING PATTERNS OF THE TIMBER TRADE

As noted earlier, Ghana's pattern of trade since the colonial period—domestic and international—and the very livelihoods of most of its people have been acknowledged as structurally problematic in the sense that they constitute economic and social imbalances that constrain the country's development (see UN/ECA, 1986). Like other countries in sub-Saharan Africa (SSA), this problem, largely seen as the consequence of SSA countries' incorporation within the global market system (Mabogunje, 1981; Sawyerr, 1990) is reflected in the pervasive primary and semi-processed commodities export bias of its economy and the high level of poverty that characterizes most of the people, of whom the forest communities represent a very significant part. Ghana's structural problems were initiated during the colonial period, when Britain organized the colony as a tributary economy to expand her own economy. Relative to the issue of decentralization of forest governance and management, Sasu (2004), for example, points out that the purpose of the decentralization exercise at the time was to solicit the support of the traditional authorities in enforcing colonial polices by providing them with a legal basis to carry out some limited local government functions such as resource management.

Ghana's pattern of trade since the colonial period—domestic and international—and the very livelihoods of most of its people are thus structurally problematic. Like other countries in Sub-Saharan Africa (SSA), this problem is reflected in the pervasive export bias of its economy based on primary products, and the associated market dependence on a few advanced countries; neglect of the domestic market; and the level of poverty that characterizes most of the people, particularly the forest communities. These attributes, which are symptomatic of underdevelopment, are largely defined, as noted by the United Nations Economic Commission for Africa, by the very structure of its economy (see UN/ECA 1986). The structural problems—the economic and social imbalances that constrain the country's development—associated with such a pattern, the causes of, and consequently, the solutions could be analyzed from two viewpoints, and this is where structural adjustment becomes the context for looking at Ghana's forest sector and the associated timber trade.

As noted earlier in chapter 4, the first viewpoint on SAP is the largely economistic perspective of proponents like the two foremost international financial institutions and sponsors of SAPs, the International Monetary Fund (IMF) and the World Bank. The second is the structuralist viewpoint of the critics including environmentalist groups such as Friends of the Earth, and the Forests and European Union Resource Network (FERN). Such groups are largely concerned with the nature of the process whereby economic growth and development is achieved. Accordingly, they focus on such issues as colonial history, social justice, and indigenous vis-à-vis external control of national destiny. From the perspective of the World Bank and the IMF, the primary promoters and sponsors of Ghana's ERP, however, market distortion was identified in the 1970s and 1980s as the root cause of Ghana's structural problems, including the declining timber industry and the associated trade. To them the decline, as noted earlier, was a condition which resulted from their perceived unnecessary state intervention in the economy. Consequently, SAPs, as a set of free-market based economic policies and prescriptions by those institutions, were therefore required to "get the prices right" in the ailing economy, such as Ghana's, with its distorted markets (Younger, 1989), and to forge an accelerated and sustainable development in the country. Generally, policy measures under SAPs involved, *inter alia*, shifting the pattern of public expenditure to the export sector; liberalizing the economy to allow the free forces of the market to determine prices; promoting exports to generate foreign exchange through an incentive system; and promoting foreign private investment through the promulgation of investment codes with extensive concessions to foreign investors (Sawyerr, 1990, pp. 20–23). Thus, SAPs basically entailed rolling back the interventionist role of the state, or the retreat of government activism in favor of a free and competitive market system and the creation of an enabling environment for investments by local and foreign capital. These measures, according to the external sponsors of the program—international capital—were basic to the restoration of the necessary conditions for long-run stable economic growth so that market mechanisms could at last come to equilibrium (see Schwarz, 1990). *Inter alia*, ostensibly, to redress the past development policy failures that cumulated in a serious economic malaise in Ghana in the 1970s and the early 1980s, which brought the country's timber trade almost to a grinding halt (see figure 6.1), the PNDC government of Ghana, at the behest of the IMF and the World Bank, adopted and implemented an orthodox SAP in April 1983.

Ghana's Pre-SAP Timber Trade

In the pre-SAP years, the traditional market for Ghana's wood exports, Europe, received at least 50% of the total volume timber produced in Ghana annually. This meant that there was not much of a shortage of timber on the domestic market. A significant feature here in terms of the trends in international trade was that about 80% of the wood exports, which was dominated by round logs and accounted for some 74% of the total, went to Europe at the time. The remainder comprised lumber, and veneer and plywood, which were 23% and 3% respectively of the annual exports. The logs were sent primarily, rank-wise, to Italy, the then West Germany, the Netherlands, and the United Kingdom. Minor markets for logs included France, Spain, Yugoslavia, Belgium-Luxembourg, the United Arab Republic, and Japan. The lumber market on the other hand comprised mainly the United Kingdom, the Netherlands, Italy, and Ireland in that order, and the minor markets included New Zealand, the Republic of South Africa, and the United States. Because a very small proportion of the wood exported was processed before 1980, timber exports therefore did not generate much foreign exchange. In 1977, for example, 1.3 million m^3 of wood from the sector were exported, fetching US $136 million. As illustrated in figure 6.1, which shows the composition of and the trend in Ghana's wood exports from1973 to 1993, there was generally a rapidly declining trend in exports reflecting the decline of the forestry sector, and indicates a shortfall in wood supply on the international market, that needed to be corrected to ensure an increased and sustained cheap supply. By 1982, the volume and income figures had declined to 100,303 m^3 and US $15.3 million respectively.

The push for Ghana's SAP by these Bretton-Woods institutions rested primarily on the premise that Ghana's economic malaise was primarily internally generated and sustained (World Bank 1981; Steel 1988). According to them, as noted earlier, problems in the timber sector and for that matter, the timber trade had resulted primarily from market distortions engendered by government interventions in the economy by way of interfering with the "invisible hand" of the market. Consequently, the solution to these institutions had to begin from within, using domestically produced resources for which the economically stagnant country has a comparative advantage (as in the production and export of timber), to engage in international trade. It was from this perspective that the World Bank (1984, p. xx) pointed out to the PNDC government that "the forestry sector offers the greatest immediate potential for growth and foreign exchange earnings" at the beginning of the program and further infused the sector with the second highest sector adjustment loan of US $157 million, out of the initial loans advanced by the World Bank to the PNDC government (Ofei-Nkansah, 1991, p. 229) to enable the country to exploit and export its timber resources.

Post-SAP Trends

With the adoption of the ERP in early 1983, as shown in figure 7.1, which shows exports of lumber and other processed wood through 2006, the end of our study period, Ghana's timber trade dramatically changed with respect to the trends in the volume, content, and direction. In terms of volume, the post-ERP years witnessed an impressive upturn and expansion in timber exports,

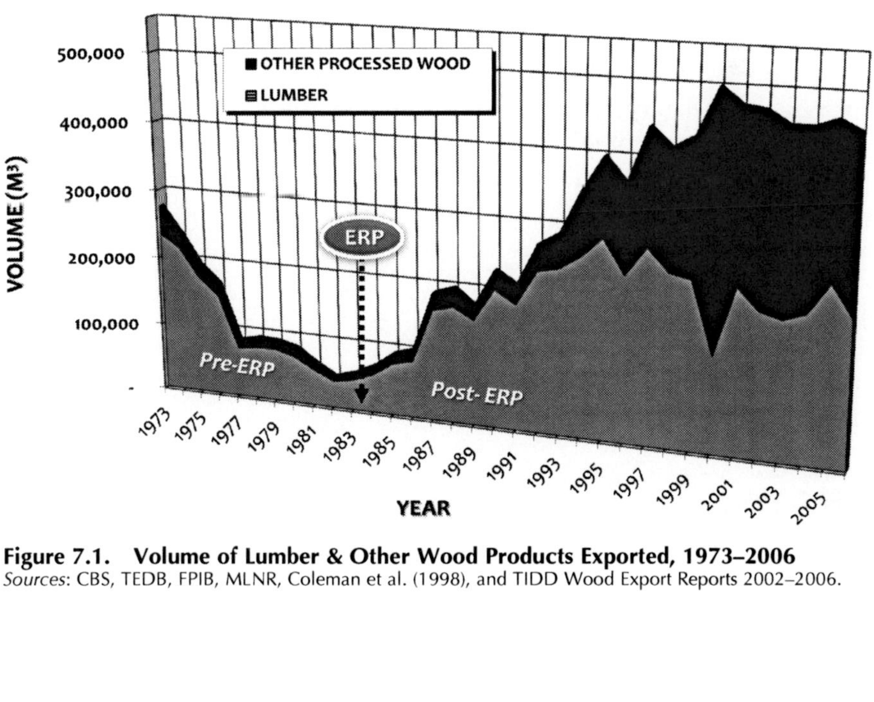

Figure 7.1. Volume of Lumber & Other Wood Products Exported, 1973–2006
Sources: CBS, TEDB, FPIB, MLNR, Coleman et al. (1998), and TIDD Wood Export Reports 2002–2006.

particularly in processed wood, and more so with the phasing out of log exports beginning in 1987. In terms of markets, Europe continued to dominate with an average consumption of about 89.9% of wood and wood products exports between 1984 and 1990, as shown in table 7.1. Regular post-ERP markets besides the traditional European market include the United States, Japan, Malaysia, Saudi Arabia, Australia and New Zealand, India, South Korea, and Libya. Relatively new markets include Greece, Turkey, Mexico, Taiwan, Kuwait, United Arab Emirates, Qatar, Sweden, and Finland. In recent times, the European domination of the export market has been waning. According to a recent ITTO's Tropical Timber Market Report (Volume 11, Number 3, pp.1–15 February, 2006), even though the EU, particularly Italy, France, Germany, the U.K., Belgium, Spain, Ireland, and the Netherlands continued to remain Ghana's main trading partner, its share has shrunk. The EU, in 2005, accounted for €92.85 million (216,840 m^3) just 50% in value and 47% in volume of the total wood exports. This was followed by the U.S.A., which accounted for 15.6% and 15% of the total export value and volume as compared with 14% and 16.47% respectively for the previous year, and significantly, the U.S.A. also represented the most lucrative market for Ghana's plywood and veneer exports.

In West Africa, the new markets include Niger, Nigeria, Senegal, Sudan, and the Gambia. There has been an expansion of the Togo and Burkina Faso markets and the pre-ERP smuggling along the Côte d'Ivoire border has ceased. All these trends have had implications not only for the environment in the form of deforestation as we saw in chapter 6, but also with respect to the traditional interdependent relationships that existed between the formal and informal sectors of the wood industry in terms of spatial integration; the domestic market for timber; the very livelihood of the forest communities; and forest management in Ghana.

Table 7.1. Volumes and Proportions of Wood and Wood Products Exported to EEC Countries Relative to Total Exports (1984–1990)

Year	*(1) Total Volume of Exports (m^3)*	*(2) Volume of Exports to EEC Countries (m^3)*	*(2) as a % of (1)*
1984	140,231	122,624	87.4%
1985	221,463	202,881	91.6%
1986	295,954	281,636	95.2%
1987	485,565	450,863	92.9%
1988	536,213	503,831	94.0%
1989	375,769	326,927	87.0%
1990	427,384	348,887	81.6%

Sources: TEDB (1990), The Ghana Timber Industry: Basic Information, Facts and Figures. Takoradi: TEDB.

ELEMENTS OF THE POST-ADJUSTMENT EXPANDED OUTPUT AND EXPORTS

As shown in figure 7.1, the dramatic expansion in the industry's export output continued to be dominated by lumber vis à vis downstream products like furniture parts, plywood, and veneer production, even though these also all recorded very significant increases. The 1993 volume of lumber exported, for example, reflected a 503% increase above the 1983 figure. The expansion of export-oriented output is also reflected in the rate of capacity utilization of the installed equipment, which shows a tremendous increase after 1984 (CBS, 6, 1983, p. 10). In the pre-adjustment years, as shown in table 7.2, the rates for all manufacturing and the wood processing sector had fallen steadily from 40.4% and 36%, respectively, in 1978 to 21.0% and 20.0% respectively in 1982. However, the rate for wood-processing began to rise steadily after the adjustment program was initiated in 1983, through 1987, and then stabilized at 70% for three years and 65% thereafter, relative to figures below 46% for the manufacturing sector as a whole over the period.

Significantly, the major increases in the rate of capacity utilisation in the wood processing sector appeared after 1986, after new investment in the form of fresh equipment and machinery imported during the 1985–1986 programs had

Table 7.2. Estimated Rates of Capacity Utilization: Large & Medium Scale Factories (%), 1978–1993

Year	*Wood Processing Sector*	*All Manufacturing Industries*
1978	36.0	40.4
1979	36.4	33.1
1980	27.3	25.5
1981	n/a	n/a
1982	20.0	21.0
1983	20.0	30.0
1984	28.1	18.0
1985	32.5	25.0
1986	n/a	25.0
1987	43.0	35.0
1988	70.0	40.0
1989	70.0	40.6
1990	70.0	39.8
1991	65.0	40.5
1992	65.0	44.5
1993	65.0	45.7

Source of Data: Statistical Service. *Quarterly Digest of Statistics*, June 1989a (p.10) and March 1995 (p.12); CBS. Economic Survey, 1982 (p. 6).

been installed. The average rate for all the firms surveyed in this study was 50%. These firms also indicated that there had been significant increases after 1983 even though they still had not been able to attain their respective 1978 levels. The steady increase in the rates of capacity utilization within the industry (and in fact, all manufacturing) after 1983 bears testimony to the fact that changes in the macro-economic conditions under the ERP were significant in the operations of wood-processing firms as discussed in chapter 5, as we looked at capital, labor, and institutional and other enabling factors that promoted increases in production under the program.

Post-ERP Changes in the Forestry Sector: Implications of the New Trends

As a renewable resource-based sector with immense potential for upstream and downstream linkages, and which "offered the greatest immediate potential for growth and foreign exchange earnings," as noted by the World Bank, the forestry sector had historically accounted for 5–6% of Ghana's total gross domestic product (GDP) and 11% of total commodity export earnings prior to the ERP. Wood from this sector had also been the third most important item after cocoa and minerals among commodity exports. In terms of volume, as shown in figure 7.1, the post-ERP years witnessed an impressive upturn and expansion in timber exports, particularly in processed wood, and more so with the phasing out of log exports beginning in 1987. By June, 1990, the PNDC government had been able to pay off all remaining recorded external payments arrears, thereby satisfying international capital in terms of arrears. The PNDC government was thus "successful" in getting back into the "good books" of international capital, and normal relations were restored with Ghana's creditors, particularly for trade financing in 1990 (World Bank, 1991), to the extent that Ghana was even touted subsequently as a "star pupil of adjustment" in international financial circles based on her "successes." All of these trends had implications not only for the environment and deforestation, but also with respect to the traditional interdependent relationships that existed between the formal and informal sectors of the wood industry, livelihoods of forest communities, and forest management in Ghana. As a result of Ghana's quest to sustain its wood exports it, *in alia*, created a domestic lumber shortage; constricted the traditional local forward linkage between sawmills and downstream processors; and further marginalized forest communities.

In order to appreciate the implications of the new trends, it is important to put the dualistic (formal and informal) structure of the forest sector in perspective, with a particular focus on the informal sector, which accommodates a large part of Ghana's forest communities (see figure 1.3). In general, the colonial and post-colonial governments' attitude to, and treatment of, the informal sector have been based on the perception of that sector as a "*diversion* of economic transaction beyond the reach of the state" and, hence, "unregulated and non-legal rather than legal" (Hutchful, 2000, cited in Sarfo-Mensah, 2005; Domson and Vlosky, 2007), which is rather contemptuous, given the important role played by the sector. It is this perspective that makes it not only imperative, but also cogent and expedient for a fundamental understanding of the sector to put it in a historical perspective. As noted earlier from the structuralist perspective by Mabogunje (1981) and Sawyerr (1990), the historical structure of relations between Europe and Africa is rooted in the diffusion of international capitalism during the colonial era as part of the space creation process of capitalism (see chapter 1). This structure was expressed in the form of unequal exchange in the area of trade, economic dependence, external control, and an unequal set of social arrangements within these spaces. As pointed out earlier in our discussion, these were all new to Africa because pre-capitalist African societies were different from the capitalist European societies. It is in this respect that the FERN Report (2006, p. 16) points out that "For centuries before colonialism, these cultures depended on forest resources. Economic production, social organization, religious belief and identity all depended on this relationship. Destruction of the forests and community exclusion from access to them, have massively disrupted these communities."

Unlike the amoral economy—the impersonal marketplace which is concerned with economic transactions between members of the society—associated with capitalism, Africa's pre-capitalist societies were essentially moral economies. Exchanges thus defined personal relationships and social relations between members of the society. Thus, prior to the diffusion of international capitalism during the colonial era, Africa's pre-capitalist economies were characterized by an "economy of affection" in which economic action tended not to be motivated by individual profit alone, but was embedded in a range of social considerations that allow for redistribution of opportunities and benefits in a manner which is impossible in modern capitalism in which formalized state action dominates the process of redistribution (Hyden, 1980). This in a way explains why policy makers, generally steeped in modern capitalism by virtue of their orthodox Western education and its associated perspectives, find it very hard to appreciate the nature and dynamics of the informal sector.

As we saw in chapter 1 and illustrated in figure 1.3, the process of incorporating the colonies into the international capitalist system led to the emergence of hybrid developing economies comprising a mixture of capitalist and pre-capitalist economies because the Ghanaian economy, like other African economies, to date, is not fully incorporated into the international capitalist system.

Ghana's economy is therefore segmented into a continuum of two relational spaces, namely the formal and informal sectors, with the formal sector being relatively economically more advanced, and involving all activities that are considered as part of a country's official production, measured by the GDP. The informal sector, on the other hand, encapsulates all other economic activities—production of goods and services outside the formal sector—and thus, operates largely outside the mainstream of government activity, regulation, and benefit. Generally then, work in this sector is often done for no cash payment or more evidently in our case, when cash is involved, it is not reported to the government as taxable income. This dualistic nature of the economy is reflected in the spatial structure of the wood processing industry. The spatial structure of an industry refers to the *location* of the industry's related elements in space or a defined spatial unit, and *all* the interactions and relationships that link those elements together as a functional unit. The spatial structure of Ghana's wood industry, with its roots in the colonial period, thus covers such components as the sources of the timber, forms of production such as logging and milling, the distribution networks, labor and markets—domestic and international.

SPATIAL INTEGRATION AND THE STRUCTURE OF GHANA'S WOOD INDUSTRY IN THE PRE-ERP ERA

In chapter 1, we noted that a spatially integrated economy has a certain wholeness to it which enables policies applied to any part of it to have system-wide effects or repercussions. Such an economy should be intrinsically centered on itself and it is not a composite of parts, some of which are autochthonous while others are more effectively involved in another and exterior spatial system, and that a colonial economy lacks this wholeness (Mabogunje, 1981, p. 62). Such spatial integration thus "involves the coordination of spatial activities to achieve a harmonious effect" (Mabogunje, 1981, p. 60). Further, Dudley Seers emphasizes the elimination of poverty, unemployment, and inequality and the promotion of self-reliance as concomitants of spatial integration. The crucial targets of such self-reliance include, he noted, "ownership as well as output in the leading economic sectors; consumption patterns that economize on foreign exchange" (Seers, 1977, p. 5) as expressed in the Government of Ghana's Report for the Sixth meeting of the consultative Group for Ghana. Furthermore, Warren (1973, p. 17) pointed out that such development should be based on the home market; that development should encompass a wide range of industries; that development should not be reliant on foreign funds; and that analogously to having a diversified industrial structure, there should be "independent technological progress" (Warren, 1973, p. 17).

At the time of independence, however, the Ghanaian economy, among other things, was largely characterized by foreign dependence, compounded by weak inter- and intra-sectoral linkages (Darkoh, 1971, 1973), described as " perhaps *the very essence of underdevelopment*" (Szereszewki, 1966, pp. 66; emphasis mine). This was mainly because "the dominating structural feature of the economy . . . was the relative unimportance of flows inside the productive system. The sectors have little dependence upon each other; most of their output goes directly to final uses" (Darkoh, 1973, pp. 10–11). In terms of downstream processing of primary products such as timber, it was also observed that, in the pre-adjustment era, firms were:

> seldom integrated in the national economy in a manner calculated to set off chains of growth in any significant direction. The utilization of by-products and waste material . . . (was not) a sizeable industrial activity; backward and forward linkages (had) only a limited role . . . and (had) been confined mainly to simple backward linkages. . . . This (could) be seen in the scale of dependence on imports for intermediate and capital goods (Darkoh, 1973, p. 9).

According to Combé (1972, pp. 4–5), roughly one-third of the foreign exchange generated by the export of timber products was used on importing wood-based materials. The reason, he noted, was that production costs were so high for extraction and processing that benefits for formal wood-based firms could only be realized through sales on the international market. A large proportion of the lumber processed was, therefore, exported as raw material, just like unprocessed logs, for international consumption. This international lumber supply was channeled through a government monopsonist agency, the Ghana Timber Marketing Board, which periodically fixed the minimum prices of wood exports. From 1963, however, the Board opened the trade to all producers and resident (expatriate) buyers, and allowed them to make their own procurement and transportation arrangements. The producers and buyers could then export the wood or sell on the domestic market. This policy reinforced the export of most of the high-grade wood. High quality wood became available on the domestic market only on special order, and at exorbitant prices. What the sawmills, which are all formal sector firms, sold as "local grade" were simply those wood products that did not meet the qualitative or aesthetic standards of the international market. This rejected timber nevertheless commanded fairly high prices due to an artificial price structure resulting from the dependence on the lucrative export trade. In addition, the supply of lumber for construction on the domestic market was irregular because it represented only the surplus from the export trade, the smaller quantities of lumber generated by some small-scale sawmills that produced exclusively for the local market, and by the informal sector (see Combé, 1972, p. 47).

PRE-ERP PRODUCTION AND MARKET STRUCTURE OF THE WOOD-PROCESSING SECTOR AND SPATIAL INTEGRATION

Figure 7.2a is an illustration of the pre-adjustment intra-sectoral linkage structures of the wood processing industry and their associated markets. The sector and the domestic market, as per figure 7.2a, are segmented into two categories, formal and informal, to indicate that the informal sector meshes into the formal at the low end of a continuum that constitutes the national economy. A broken line is used to reflect the view that inter-segmental movements of factors of production and output are possible, in fact, frequent in the case of labor and goods (see Moser 1978; Ninsin, 1991; Meagher, 1995; Yeboah, 1998). Two types of arrows are used to represent the production and distribution flows within the wood processing sector as a whole, namely the large arrows (for linkages) and the thin arrows (for product destination or market). Official data for the actual volumes are generally unavailable, hence the arrow-types are for purposes of differentiation (their respective sizes do

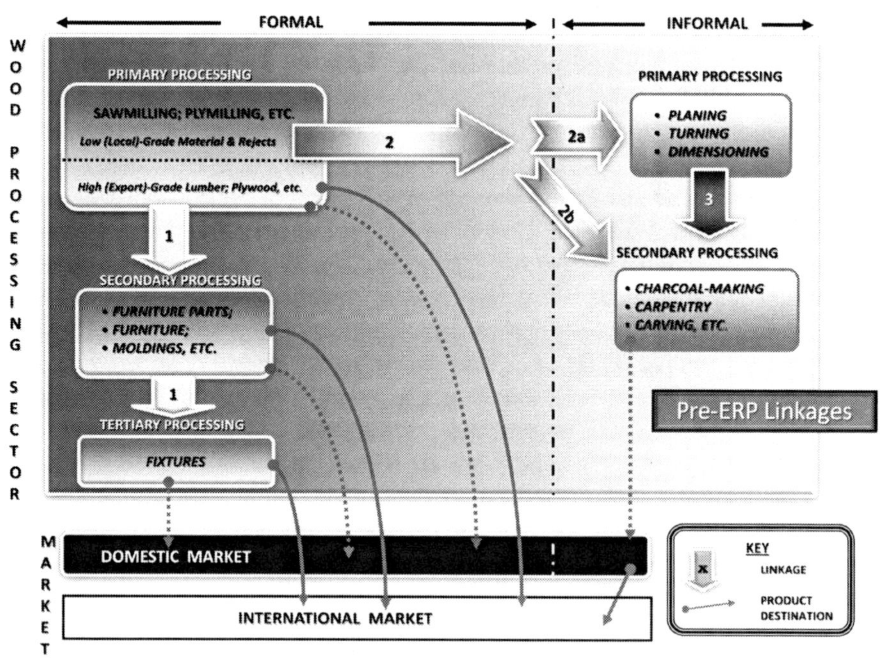

Figure 7.2a. Pre-ERP Linkages between Ghana's Formal and Informal Wood Processing Sectors and Associated Markets.
Source: Owusu (2001c)

not indicate the volume of flow). The large arrows represent the three types of sectoral linkages identifiable within the industry, namely the *intra-formal* (1), the *formal-informal* (2), and the *intra-informal* (3) sectoral linkages. The *intra-formal* sectoral linkage system shows the established traditional system of forward linkages from the primary processing stage through the tertiary stage in the pre-adjustment era. Residual lumber from the export-oriented sawmills, and regular lumber from the small-scale sawmills for domestic consumption, served as inputs for secondary processing such as furniture manufacturing. Further downstream, a small proportion of products from secondary processing, excluding the leakage into the domestic and international markets, was used as inputs for tertiary processing (e.g., floorings) in vertically integrated and non-integrated firms (see Combé, 1972). A notable feature of the limited intra-sectoral linkages was the problem of waste, with wood recovery rates averaging only 42% (Chryssides, 1974). The more integrated a firm was, the higher the recovery rate from its wood inputs. Smaller pieces of wood or "off-cuts" were converted into such items as wooden crates, moldings and floorings.

The formal-informal sectoral linkage system (2) shows that the relations between the two sectors were via the direct sale of local (low)-grade or unwanted materials as inputs from the formal sector to the informal. As shown by arrows 2a and 2b, the sawmills in the formal sector sold their "off-cuts" from logs, "rejects" from export parcels, and other low-grade material to the informal sector. Some of these were then planed and resized for sale, or used by small-scale artisans such as carpenters and traditional wood-carvers as inputs for small-scale construction, "roadside furniture" (for low-income groups), and items like traditional handicrafts or artifacts (e.g., carved wooden dolls, stools, and masks). Some of the "rejects" were used as fuel wood or charcoal for household and other uses. These activities straddle the formal and informal sectors.

The major markets of the informal sector are found in Accra, Kumasi, and Takoradi, the top three cities in the country's urban hierarchy, and their nature has not changed significantly from the time of adjustment (see Boateng, 1991, p. 14). Products from these markets are also found throughout Ghana, and some of them find their way to other West African countries such as Burkina Faso, Niger, Mali, Togo, and Senegal (Boateng, 1991, p. 14). The production activities described above also show a *traditional* intra-informal sectoral forward linkage system (arrow 3), between primary and secondary processing in that sector. There was no tertiary processing in the informal sector since the types of equipment required for such products as kiln-dried furniture parts, lippings, and floorings are too sophisticated and expensive for informal sector operators. Formal wood processing had substantial inter-sectoral linkages with the construction and transport sectors, but the linkage with the energy sector was rather limited (see Combé, 1972, pp.111–12). On the whole, therefore, the pre-adjustment intra-national linkages to the wood processing industry were generally weak (see Darkoh, 1973, pp. 9–11; Huq, 1989, p. 128). Instead, the formal sector of the industry produced mainly for the external market (see Mabogunje, 1981), while it relied mainly on the informal sector to meet the demands of the domestic market.

Post-ERP Production and Market Structure of the Wood-Processing Sector and Spatial Integration

In contrast with the pre-adjustment era, post-adjustment developments, as illustrated in figure 7.2b, suggest that the most important and significant change involves the *intra-formal* sectoral linkage system: the traditional forward linkage (now represented by arrow 1a) between the sawmills and downstream processors in the formal sector, was largely destroyed—with expansion implications for the informal sector. This undermined the process of spatial integration as it conflicted with the fact that a region's competitive po-

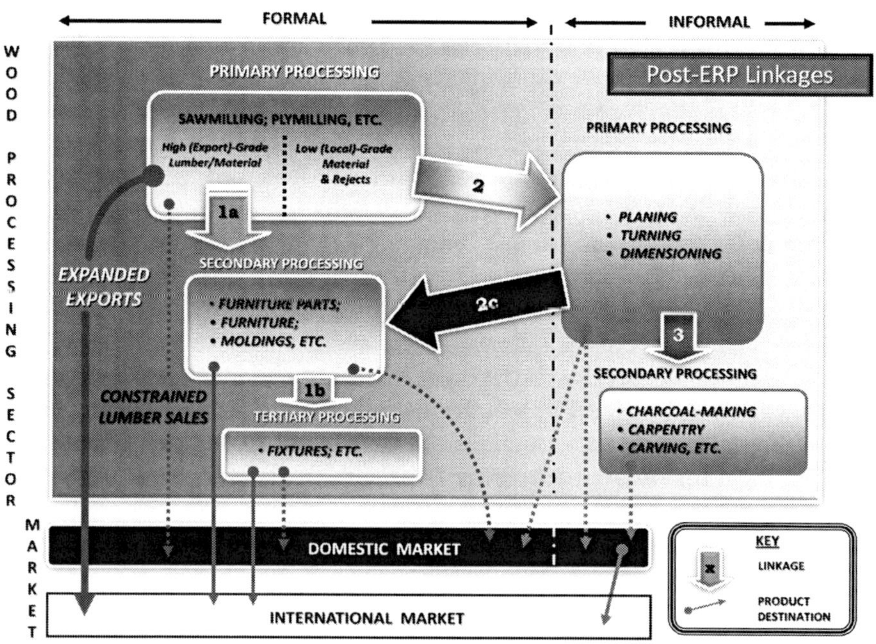

Figure 7.2b. Post-ERP Linkages between Ghana's Formal and Informal Wood Processing Sectors and Associated Markets.
Source: Owusu (2001c).

sition and relative growth potential is directly proportional to the density and strength of linkages found within its production network. This is because, as discussed in chapter 1, generally the stronger and more numerous the linkages within a region, both intra- and inter-sectorally, the more likely that region is to experience and sustain growth, since flows and linkages are largely responsible for determining a region's economic status (Fik, 1999, pp. 20–21).

This unprecedented change in the *intra-formal* sectoral linkage system was largely due to the dramatic expansion in lumber exports to the international market to generate foreign exchange (illustrated by the left-most product destination (thin black) arrow labeled "Expanded Exports" in figure 7.2b), thus destroying the traditional forward linkage, which is basic in the production network of any integrated economy, and which is a critical element in any regional development equation (Fik, 1999, pp. 20–21). Thus it is this destruction of the traditional forward linkage system that generated a major shortfall on the domestic market. This expansion in lumber exports has in the process constrained lumber sales on the domestic market; as observed by Barker et al. (1991, p.1), "the millers are not supplying local manufacturers, and the manufacturers are not able to produce the quality or quantity of goods the export market demands." This was exacerbated later by the imposition of a levy on air-dried lumber. In examining lumber as the dominant forest product exported in the post adjustment years, Donkor (2003, pp. 69–70; pp. 118–19) noted that air-dried lumber initially constituted the main component but was overtaken by kiln-dried lumber in 1997. This was because in 1996, the government imposed a levy that ranged from 10–30% FOB depending on the rarity of the species, on air-dried lumber. This levy—a directive from the government to the then Forest Products Inspection Bureau, now part of the Timber Industry Development Division (TIDD) of the Forestry Commission—served as a disincentive to primary processing by reducing the formal production of air-dried lumber and increased kiln-dried lumber.

This domestic lumber shortage situation led to an emergence of an intensive *formal-informal* sectoral linkage system (the arrow 2 group). Arrow 2c illustrates the associated general shift of lumber-sourcing for secondary stage processing from the formal sector to the informal sector, where prices for the re-processed "off-cuts" and rejected lumber, previously purchased from the sawmills were lower. Thus, some formal sector firms such as producers of furniture, furniture parts, and fixtures, generally procured low-grade material from the informal market, largely supplied by chainsaw operators, as inputs instead of the high-cost lumber from sawmills in the formal sector. As a result of the expansion in lumber exports, therefore the informal sector has become dominant in meeting the needs of the local market (Birikorang et al., 2001). This is noteworthy because the informal sector is a "struggling" entity, which is generally neglected and discriminated against by the government (see UN/ECA 1986, p.3; World Bank 1989, p.135), and thus excluded from all the ERP-related production incentives. It must be noted that as part of the creation of an enabling environment for investments by local and foreign capital under the program, the wood processing firms were given incentive packages under the investment code (PNDC Law 116, 1985) to encourage exports. This also added to the problem since they discouraged sale of lumber on the domestic market. The sawmill owners and managers preferred to enjoy tax rebates based on volumes exported. Additionally, they could retain at least 35% of the export earnings in an external account (for safe keeping), as well as an extra portion of the earnings ostensibly for the purchase of machinery, equipment, and spare parts. Because it was convenient for the government to have the informal sector—through the chainsaw operators—meet the local demand as it (the government) concentrated on exports to accommodate international capital, it at that time deemed it proper to issue the "now-illegal" chainsaw operators permits to operate. As Odoom (2005, p. 15) puts it, "The ERP led to the legalization of chainsawing as a result of increased demand for lumber to support the dramatic growth in construction activities."

It is therefore the informal sector, through the chainsaw operations, that stepped in to bail out the government's own formal sector by providing the 70–85% of the lumber needs of the domestic market (TBI-Ghana, 2003). According to a Forestry Research Programme (FRP) *Country Report on Ghana*—(R 8509), the formal sector or the "traditional mills by law (Section 36 of L.I. 1649), are obliged to supply not less than 20% of their mill output to the local market. If this law is complied with, the mills will deliver just about 120,000 m^3 to the local market. This will represent only 24% of the estimated local consumption of 500,000 m^3" and it is the "difference between the legal supply and estimated demand [that] is filled by illegal chainsaw lumber" (Coleman 2004 cited in FRP, 2006). In fact, the domestic demand is likely to rise or to keep pace with the expansion in Ghana's expanding building construction industry and the growth of the economy (Domson et al., 2007). This must be viewed against the backdrop of the fact that the sawmills are characterized by an overcapacity of outdated and inefficient equipment (ITTO, 2005). Such processing overcapacities contribute to illegal logging and other predatory forestry-related behavior (IUCN, 2006), and there is evidence that "this overcapacity is increasing due largely to new investments to take advantage of the relatively cheap raw material and the existing loopholes in investment incentive provisions" as noted by Domson et al (2007, p.10).

It is in this regard that Ninsin (1991) points out that under free market conditions, contractual relations emerge to bind informal sector operators to the formal sector, either for the supply of raw materials or semi-finished goods to the latter. That aside, it is really unrealistic to expect the poor farmers and other poor rural people in villages and the small towns in the forest region—the forest community—to travel to the timber markets or sawmills in the big cities or district centers to buy lumber from the formal sector for their own private construction needs, at formal sector prices and incur transportation costs, when chainsaw operators are more easily accessible, and can deliver their needs at a much lower and affordable price. As a matter of fact, as noted in the FRP Report (2006), the formal sector sawmills cannot even sell their products on the domestic market at prices commensurate with their production costs; consequently, they are compelled to export their products to enable them to stay in business, thus intensifying the international forward linkage with overseas customers who further process the wood. It is therefore in this respect that the informal chainsaw operations, deemed as "illegal," become a critical component in the Ghanaian market for wood, more so as the operators seek to create, expand, and defend their operational turfs or space against the government's formal sector operations, thus making their turfs a contested space.

THE SPATIAL STRUCTURE AND DEVELOPMENT OF CHAINSAW OPERATIONS IN THE WOOD SECTOR

As a contested space or operational turf relative to the functional link between space and process, actors in the formal sector therefore generally perceive and treat the informal sector activities such as the chainsaw operations as "illegal" and need to be banned since they are often beyond the reach of government to be regulated, as are the incomes thereby generated be directly accessed and appropriated through taxation. It is in this regard that the informal sector is described as "a diversion of economic transactions beyond the reach or boundaries of state and formal economic structures" (see Hutchful, 2002; Sarfo-Mensah, 2005; Domson & Vlosky, 2007; and Birikorang et al., 2007). This situation, as we shall soon see with regard to the informal sector's chainsaw operators—loggers and millers, the main primary producers in the domestic timber trade—is tied to a form of turf conflict between the actors in the two sectors. This is because, as with any socially created turf, their respective turfs tend to be formally or informally actively controlled, preserved, protected, and expanded by the related actors, if expedient, with implications for the related stakeholders. This is because turfs as relational spaces representing related actors' ambits of power, tend to be either formally or informally actively challenged, contested, or encroached upon by other groups.

Production of timber and other wood products for the export market has been the dominant orientation, in terms of policy and encouragement, of both the colonial and post-colonial governments, with very little attention to production to meet the growing demand on the domestic market. This has been more so in the immediate pre- and post-ERP periods. This was largely because the construction sector, particularly roads and housing, which consumes much of the timber and wood products was also in crisis like other sectors of the economy. As noted by Sarfo-Mensah (2005), it had sunk below the threshold of minimal capacity utilization, and consequently there was no urge for the government to even consider any policy on meeting demand on the local market. In fact the government's pre-occupation now was to accommodate international capital by focusing on timber and other raw material exports to address its huge balance of payment deficit, as we saw in chapter 6. There was thus no coherent policy of meeting the largely ad hoc domestic demand for timber. It is in this sense that Ninsin (1991) points out that the informal sector renders useful and productive functions in African economies without formal controls and compensation. That the informal sector provides useful and productive functions in the Ghanaian economy in this respect without formal controls and compensation, is illustrated by the fact that much of the domestic supply of lumber is provided through the so-called "illegal" chainsaw logging and milling. As shown in *The Ghana Wood Industry and Log Ban Export Study* by Birikorang et al. (2001) quoted in Bird et al. (2006, p.5), it is estimated that in 1999, out of the estimated 3.7 million m^3 of timber harvested, illegal chainsaw activities accounted for as much as 46% (1.7 million m^3). What then is the nature of chainsaw logging and milling, and how did they emerge as actors in the timber supply chain? Who are the people engaged in such timber production? Why do they do it; and what makes chainsaw operations illegal?

According to the Forestry Research Program (FRP) Country Report—Ghana (R-8509), *Chain Saw Logging and Milling in Ghana: Background Study Report*, chainsaw machines were introduced to Ghana in the 1960s. They improved felling of trees during logging. The use of the machine was revolutionary in the sense that it increased felled volume (per man per day) by 100–400% over the manual two-man hand saw technique for felling and bucking. They replaced the long manual blades for felling and cross cutting trees. Subsequently, the chainsaw machine operators were introduced to the techniques of converting logs to lumber using the free hand method, particularly in the 1970s. Since then, chainsaw milling has replaced pit-sawing for the supply of household lumber, thus ushering in a new informal sector, as such logging and milling became an important enterprise among the deprived rural communities in the forest zone. The chainsaw logging and milling enterprise and the associated trade is a very significant part of the overall timber sector, involving many functionally linked actors who may be individuals or part of small informal groups. Marfo (2004, p. 25) describes the organization of the chainsaw business as a network of a number of actors whose agency is enrolled by the chainsaw operatives, namely all actors directly involved in using or causing chainsaws to be used to access timber and convert it into lumber. This involves (a) the timber sellers in towns and cities; (b) the chainsaw operators who own and use chainsaws to cut and convert timber in the forests; and (c) the forest or rural folk who are either private business people or people hired by timber sellers who act as middlemen between timber sellers and their teams in the forests. The chainsaw loggers and millers, by virtue of the informality of this subsector of the timber industry, do not receive any formal or organized skill training in the business. Rather, they learn the trade from other operators who have had long working experience in the business.

In terms of the spatial structure of the chainsaw timber operations, the operators source their logs, the basic raw material, from farm lands, farm fallows, and to some extent from some of the government protected forests or reserves. The mode of acquisition of logs, as noted by the FRP Report (2006), "is highly variable depending on the social norms relating to tree and land tenure of the people in the community and may be through authorized or unauthorized means." In terms of labor or employment for the poor, and as a source of livelihood for some 50,000 people employed at various levels of field operations alone (stump site operations—felling, sawing, to head portage; transport; marketing; and utilization) as noted in section 2.0 of the FRP Report (2006), this spatial structure involves:

a. Chainsaw machine operators and their apprentices, and
   Porters who carry the planks from stump site to loading points
b. Loading gangs who on-load haulage trucks
c. Transporters (truck owners, truck drivers, and their assistants) who haul the lumber to the markets and from markets to consumers or users
d. Market hands who assist in off-loading the trucks from the forest and loading trucks to consumer delivery points
e. Lumber brokers who buy and retail the lumber from the chainsaw operators; and
f. Consumers who buy lumber from the timber markets or directly from the chainsaw operators.

The main consumers who constitute the domestic market for the lumber include small- and medium-scale construction enterprises, individual builders, small scale timber merchants, carpentry shops, and joinery producers. The FRP Report (2006) noted that there were over 41,000 small scale carpenters and

25 medium to large scale furniture and joinery firms. Items produced by the carpentry and joinery enterprises include household and office furniture, furniture for schools and other institutions, panel doors, door and window frames, windows, pallets, fruit crates, caskets, and cabinets. Domson et al. (2007) point out that as many as 6,000 tertiary enterprises are constituents of the informal sector. According to the FRP Report, these consume about 75% of the country's annual sawn timber requirement, with the remaining 25% met by the formal construction and building firms. As noted by Sarfo-Mensah (2005), the domestic market has been expanding, particularly in the post-adjustment era, with the emergence of new industries, the construction of major highways, and the rehabilitation of hospitals and educational institutions. Other items driving the demand for lumber, he notes, include truck body building, railway sleepers, and boat building. Significant quantities of household and other furniture, doors, and other such items are also purchased by Ghanaian and other West African merchants, and transported overland for sale in such countries as Burkina Faso, Niger, and Mali.

In terms of the physical location of the informal markets, all ten regional and the 110 district capital towns/centers have lumber markets. The principal ones, however are Kumasi's *Anloaga* in the Ashanti region, which covers some fifteen hectares and the biggest; *Ofanko* in Accra, *Techiman*, in the Brong-Ahafo Region, *Kasoa* in the Central Region, *Kokompe* at Takoradi in the Western Region. These markets are run by lumber brokers who source their stock principally from the chain sawyers, and to a much lesser extent from the formal sawmills (see Birikorang et al., 2001; Bird et al., 2006) as we shall see in detail soon with regard to the constriction of the traditional forward linkages in the formal sector of the industry as a result of the ERP. As a reflection of the on-going tension and conflict between the formal sector and the informal sector, despite the major contribution of the latter in the economy, the role of the chainsaw loggers and millers who contribute some 70% of the domestic market wood supplies have been ruled as illegal. This is tied to the argument by the formal sector actors that, even discounting the common felling and milling accidents and fatalities associated with the operations of chainsaw operators, their operations are very wasteful, inefficient, and very environmentally unfriendly with regard to extraction and processing. Arguments advanced for the control of such operations include a portrayal of chainsaw milling as an activity that is not conducted under any standard set of working procedures. Consequently, such uncontrolled activities result in indiscriminate exploitation of tree resources, and reduce natural environmental and ecological processes such as the protection of watersheds. The FRP Report (2006) for example suggests that illegal chainsaw operations in some forest reserves and water catchment areas are reported to have contributed to increased pollution due to uncontrolled disposal of spent oils, and a decline in water levels due to siltation of water channels. Furthermore, they note that such operations result in the destruction of habitats, and loss of biodiversity through the removal of seed trees as well as immature ones resulting in loss of regeneration.

It is estimated that about 68,000 workers are involved in illegal chainsaw timber harvesting (Oteng-Amoako et al., 2008). According to a Forestry Commission study quoted in TBI-Ghana (2003), illegal chainsaw activities alone accounted for about 1.7 million m^3, and together with other illegal logging activities and the legal harvest, the total harvest in the early 2000s was about 3.7 million m^3, which was almost double the annual allowable cut of 2 million m^3. Based on these figures, it is argued by the policy makers that lifting the ban on chainsaw operationswould even exacerbate further the deforestation problem by increasing the log harvest to 5 million m^3, and this is because the use of chainsaw for lumber production—chainsaw milling (CSM)—as well as the sale and use of chainsaw lumber in Ghana were therefore banned in 1997 (Act 547 of 1997) and 1998 (L.I. 1649). Prior to this ban, there existed a symbiotic relationship between the actors of the two sectors in the form of a 1991 legal provision, the Trees and Timber (Chainsaw Operations) Regulations, a legislative instrument under which the chainsaw operators were legally recognized, with the proviso that their activities were to be regulated by the District Assemblies and District Forestry officers at the local level. Essentially, it was largely a decentralized resource allocation management regime put in place in recognition of the important role played by the chainsaw operators in the domestic market (FRP, 2006).

LUMBER PRODUCTION AND FORMAL-INFORMAL TURF CONFLICTS

The abandonment of the symbiotic relationship therefore meant an incursion by actors or policy makers of the formal sector into the informal timber producers' turf; this interferes with the latter's activities, thus putting the chainsaw operators in a self-protection-cum-survival mode. It is this self-protection mode and associated activities that make the ban very difficult. To assert its control over the chainsaw operators turf, the government has sought to control illegal timber harvesting, especially outside the forest reserves, and have included several ad hoc control measures such as the establishment of mobile forest protection action groups and strict felling controls; coordinated actions by both the military and police to crack down on operators; and the confiscation of lumber, equipment, and vehicles used in the illegal harvesting operations (Bird et al., 2006, p. 5). For the informal timber producers—the chainsaw operators—Marfo (2004) has in fact identified some of the rather sophisticated strategies adopted, as a response to the government's attempts in the turf conflict. These are (1) the establishment of certain informal business alliances, (2) nocturnal operations, and (3) enrollment of the agency of the state. They have also developed related tactics in dealing with (i) opposition at the community level, and (ii) law enforcement personnel comprising the police and the forestry officials, as well as (iii) coping strategies on the ground. Let us now look at the nature of these strategies.

The Informal Business Alliances

These involve various informal arrangements between the timber sellers on the one hand, and the chainsaw operators on the other. There are several variants of such alliances, which may be directly or indirectly between the timber sellers and the chainsaw operators, and depending on the terms of the alliance, the chainsaw may be owned by the operator or the seller, and operational funding may also be provided by the seller or the operator himself. The first typical variant of this arrangement is one in which the seller is solely dependent on the operator for the chainsaw lumber, often referred to on the market as "bush cut." In this particular arrangement, the seller hires the operator or an agent from a community and sponsors his operations with the understanding that the sponsor becomes the monopsonist of all the lumber produced by the agent or operator. In this case, the operator or agent, as the case may be, becomes a front man in the community for the seller. The seller has a couple of advantages in this relationship: He avoids the cost of operations because the operator, being a local resident or native, negotiates down the timber price demanded by the farmers or land owners and the compensation costs of destroyed crops since the farmers are apt to charge non-natives or city-folks higher prices. The sellers thus avoid having to bribe or shell out extra monies to the chief, the assemblyman, or the forestry officials to facilitate the transaction. The seller also reduces the cost and trouble of identifying farmers, landowners, or lands with commercial timber trees. In this case, the operator or agent uses his social contacts in the community to contact the landowners or farmers and also mobilize personnel needed to produce the lumber for the seller. When the lumber is ready, the seller makes the final payments and transports the lumber to the market for sale; thus this alliance makes the whole lumber procurement process easier and faster for the seller, while the operator avoids the trouble of transporting and disposing of the lumber on the market himself, and is assured of his remuneration.

Another typical relationship is one in which the operator may be on his own, in the sense that he is responsible for all transactions and activities prior to his sale of the lumber to his clientele of sellers. In this case, therefore, the agent uses his own resources to finance the operations to get the lumber ready for sellers. He then deals with sellers who would make their own arrangements to transport the lumber to the market. The operators and the agents also tend to establish business relationships with the local actors, especially the chiefs and farmers. The farmers may sell the trees to the operators or agent, or inform them of available trees for which they negotiate and settle on mutually acceptable payments.

Such local operations are also strengthened by the income earning opportunities offered to the young unemployed people who become part of the network by serving as loading boys and other positions of assistance for income. To maintain the income opportunities, the local members of the network help as beneficiaries to sustain the chainsaw lumbering operations by refusing to cooperate with the authorities. Oftentimes, in the alliances established, the seller tries to avoid being implicated in cases of arrest by the authorities.

Nocturnal Operations

An important strategy used by the chainsaw operatives, who are often armed, is to work in the night to outmaneuver the forestry officers who mostly work in the day and are often under equipped (Tropenbos, 2003, p. 3). Nocturnal operations enable them not only to evade arrest by the authorities but also to reduce operational costs by evading tree price payments and compensation to the farmer, the landowner, and the chief as expected of them. As an armed group, this makes them dangerous to any formal or informal group that would try to halt them, because all the participants—those who do the cutting, sawing, packing, loading, and transporting—see their very livelihoods at stake, and like a gang, would do anything to protect their turf.

Enrollment of the Agency of the State

The production process becomes complete with consumption, and consumption of the chainsaw lumber is only possible if the lumber is successfully transported to the timber market in the towns, cities, or any other destination, a process in which the enrollment of the state or the co-optation of some state officials becomes critical in the chainsaw lumber trade. It is in this regard that the ability of state officials—the corrupt ones—to move from the formal into and out of the informal sector to operate when they deem it convenient or personally profitable in Africa's suspended states, as noted in chapter 1, is illustrated par excellence. To ensure that timber is formally produced for home consumption or for export, the government monitors the movement of the timber as part of law enforcement. Forestry officials, the police, and Customs,

Exercise and Preventive Service (CEPS) officials are authorized to inspect the lumber transporting vehicles and demand related permits and papers (see Harvey, 2005, p. 2). If the transporters are unable to provide valid documentation certifying the legality of the timber, the load along with the vehicle would be impounded and the operatives involved are arrested and prosecuted.

As noted by Marfo (2004), this chainsaw operatives' strategy in this formal-informal turf conflict has primarily been to develop "obligatory passage points" by escorting the chainsaw lumber-laden trucks via an accommodating or responsive "operations road." These operations roads are particular routes or areas that the operatives have conducted their business on for some time and have become familiar with the people and officials or established the necessary facilitating alliances in that area. On such operations roads, the transportation process is thus less risky, provided that they are able to negotiate and make the necessary payments or pay the obligatory bribes, and in so doing forge alliances. With the previous alliances established and payments made, the operatives tend to enjoy friendlier treatments and are able to effect new negotiations with relative ease. Outside such spaces of operations, it is more difficult to transport the load to the destination because of the uncertainties about law enforcement, which may result in such situations like confiscation and/or arrest. In very new cases or operations in unfamiliar spaces or routes, the operative travels in a car ahead of the timber laden truck and alarms the truck driver, often by cell phone of any imminent danger such as the police, CEPS, or forestry officials check points. Generally, however, as noted by Quartey (2010), "in many cases, checks of timber in transit by the Timber Industry Development Division (TIDD) are a mere formality due to corruption." Quartey sums up the situation, with the government's complicity, this way: "it goes without saying that CSM has been endorsed: as long as chainsaw millers do not get caught, the illegal timber gets approved in the market (VAT is even being charged on the timber)."

As noted in chapter 1, turfs tend to be formally or informally actively controlled, preserved, protected, and expanded by the related actors, if expedient with implications for the related stakeholders. Generally, in this formal-informal turf conflict, the operatives have developed several tactics in the quest for controlling, preserving, and protecting the chainsaw operations, for dealing with the obstacles with respect to local or community-level opposition, and dealing with law enforcement *via* the police and the forestry officials:

Dealing with Community-Level Opposition

As a way of formally and actively controlling, preserving, protecting, and expanding its turf, the government has been engaging in a campaign in criminalizing the informal chainsaw activities. Consequently, there has been a growing public concern and awareness regarding chainsaw operations that have resulted in various forms of resistance, really not to stop logging but the conveyance of the chainsaw lumber at the community level. To combat such local opposition, the primary strategy has been threats of physical assault using offensive weapons (Marfo, 2004). Some chainsaw operators do not hesitate to physically attack community-level personnel and officials when they feel their interests and livelihoods are threatened (see *Daily Graphic*, September 6, 2003, p. 22). As noted in the FRP Report on Ghana, (2006, sec. 1.4), this included vandalizing of Forestry Department's vehicles, murder, and assaults on law enforcement officers and the police. Their ability to resist the government this way, often with the connivance of some local chiefs and community leaders, indicates not only the ability of such leaders to shuttle between the formal and informal sectors for personal gain when convenient, but also a response to the government's incursion into their traditional turfs to exploit their resources without adequate compensation.

Dealing with Law Enforcement Personnel

The assistance of the military may sometimes be used in law enforcement relating to the activities of the chainsaw operatives (Bird et al., 2006, p.5; FRP Report on Ghana, 2006, section 4.0), which along with CEPS, the main law-enforcement personnel in the turf struggle, comprise the police and the forestry officials (see Harvey, 2005, p. 2). Generally, official remunerations of law enforcement personnel tend to be very low, which makes them susceptible to bribery and corruption. In dealing with the police, therefore, the chainsaw operatives' main strategy exploits the ability of such personnel to shuttle between the formal and informal sectors for personal gain when convenient, by negotiation for rent payments or bribes. It is through such links that the chainsaw lumber laden trucks are able to go through security check points. In this respect, the police engage in the negotiation using some front people in the form of interested parties like the driver or a junior ranked police officer in the team. A narrative from an informant quoted in Marfo (2004, p. 27) aptly summarizes the nature of the negotiations:

> The trick used by some policemen is that, one, usually the senior among them will instruct the others to arrest the vehicle and accompany it to the Forestry Department or police station. The other(s) will then approach the driver trying to suggest that they should find "something" for them to use to beg the officer to allow them to go. Through this, negotiation ensues and depending on the value of the lumber, and the negotiation skills of both, a deal is reached.

As an initial successful transaction, this process may establish an operations route and a business alliance. For subsequent trips therefore, payment becomes routine if the same officers are encountered. If the negotiations fail, as Marfo (2004) further points out, the vehicle may be impounded and the operatives arrested. Even under such circumstances, the chainsaw operatives may still be able to exploit the proclivities of the police by paying their way through at the police station. The operatives in certain cases enroll the agency of the police by arranging to have an influential policeman to "escort" their trucks to the destination, in which case the "passage lubrication" process is taken care of by the police through some sophisticated networking involving government officials.

In dealing with the forestry officials, the chainsaw operatives, again, exploit their low official remunerations and their ability to shuttle between the formal and informal sectors for personal gain when convenient. Marfo (2004) shows that some operatives establish alliances with the forestry officials, buy confiscated chainsaw lumber, and obtain the necessary official receipts and waybills to convey the lumber to the markets. Some of the officers even tend to use under invoicing to maintain a portion of the sale, all of which tend to establish and sustain business alliances ensuring a continuous supply of confiscated lumber for sale for profit. In situations where the operatives are unsuccessful in bribing their way through security check points, they often enter into negotiations with the forestry officials and purchase the confiscated lumber from them rather than abandoning the load, which may be later sold to other parties. The sophisticated enrollment of the agency of the state by the chainsaw operatives really does not mean that the former has totally been co-opted by the latter.

As noted in Marfo (2004), between January and June 2003, about 500 trucks of illegal chainsaw lumber were impounded at the Forestry Services Division office in Koforidua, in the Eastern Region alone. However, due to the low court fines compared with the profit of the trade, prosecution has not been effective in the state's attempt to curtail the chainsaw lumber trade. In general therefore, despite the government's strategies of control, the chainsaw operators have been able to contain the state's incursions, its laws, and apparatuses to survive the struggle (see Marfo, 2004). Birikorang (2001) quoted in Marfo (2004) observed that the policy of chainsaw ban is unenforceable, as it encourages rent-seeking behavior or corruption among the Forestry Commission staff, local government representatives, and government task forces, and that the presence of a de facto authority involving a triangular relationship and the lines of influence among the local stool, district officers, and the community hinders forest policy through chainsaw activities.

As pointed out by Bird et al. (2006, p. 5), the government's control measures have largely failed to have the desired impact mainly because of the high demand for wood—due to lumber shortage on the domestic market—and the low penalties involved. In consistency with our discussions above, the Verifor Case Study (Bird et al., 2006, p. 5) notes further that "legislation, regulations and codes of practice put in place to control harvesting and to protect the forestry resource have been either inadequate or not properly enforced." Significantly, it is pointed out that "the reasons have much to do with the political economy of the country, which dwarfs the technical reasons that are often cited." It is in this respect that we looked at the issue of lumber shortage and the relevance of the informal-formal sectors interactions as key issues in Ghana's political economy. The formal-informal sector interactions are also reflected in the marginalization of the forest communities under the program. Before discussing the marginalization of the forest communities, there is the need to recall, as was noted in chapter 1 regarding the structure of African economies, that within the continuum from the informal sector and the formal sector, there exists an inter-segmental space within which corrupt formal sector or public officials and bureaucrats can collude with informal sector operators, including traditional authorities, for mutual benefits or gains, especially since the former group tends to be largely underpaid, in a number of cases, or are just greedy, or take advantage of tempting situations for extra income, when accountability to the state or state oversight is minimal or nil.

THE FORMAL-INFORMAL SECTOR CORRUPTION LINKS

In examining the issue of corruption in Ghana's timber industry, Arthur et al. (2007, pp. 31–33), identified three HF decay situations involving corruption sequences in the formal-informal inter-sectoral links and the associated power relations among the actors. These sequences are subsumed under the following corruption cycles: a) National Power; b) Traditional Authorities; and c) District Forestry Officials, as discussed below.

The National Power Corruption Cycle

This cycle is couched in the view that the state apparatus has been conditioned to undermine the communal or domestic economy and support the growth of corporate capitalism in all spheres of life, since the colonial era—essentially, the accommodation of international capital argument (Sawyer, 1990; FERN, 2006).

Figure 7.3 illustrates this self-sustaining corruption cycle linking national actors—the key decision-makers, the politicians, and technocrats—and the so-called "Timbermen." As a group, these Timbermen represent both international

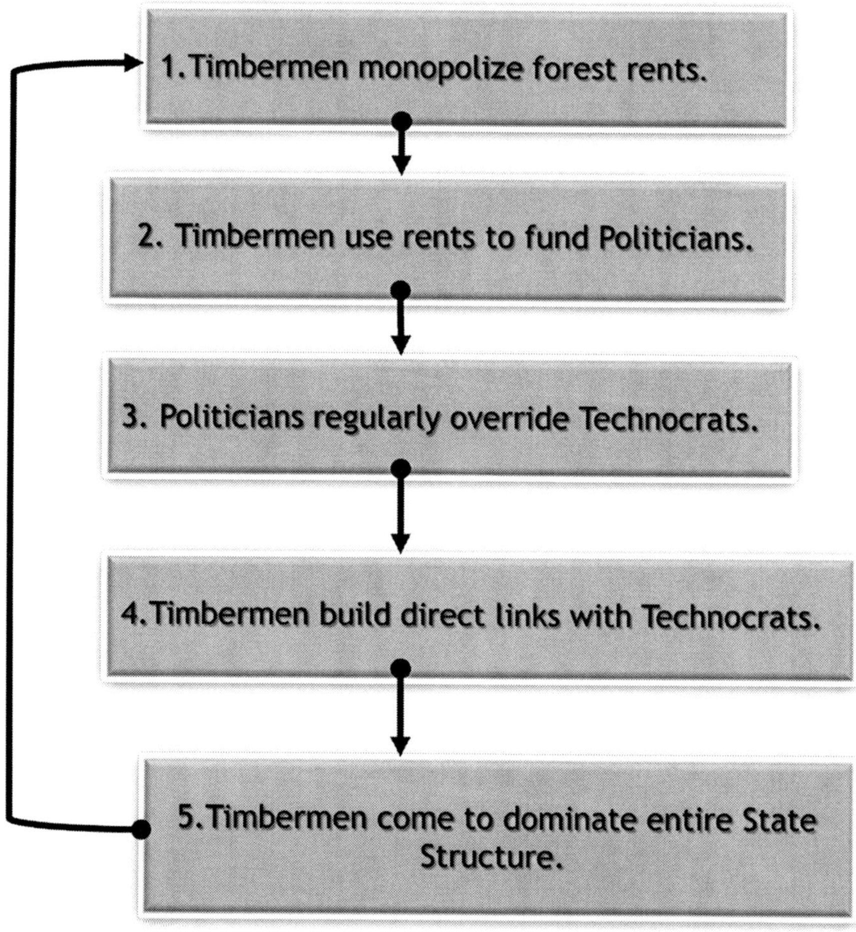

Figure 7.3. Corruption and Power Relations among National Actors and Timbermen
Source: Adapted from Arthur et al. (2007).

and domestic capital. This group comprises the owners of the timber companies, the so-called "timber contractors," concessionaires, and their respective representatives and agents. This linkage between the national actors and the Timbermen inevitably undermines national policies relating to the timber industry in general, since it accords the Timbermen a tremendous power and influence at the national level and sustains their power to monopolize forest rents.

The Traditional Authorities Corruption Cycle

Figure 7.4 links the Timbermen with the Traditional Authorities (TAs)—the traditional custodians of community lands, such as the Chiefs, who are cus-

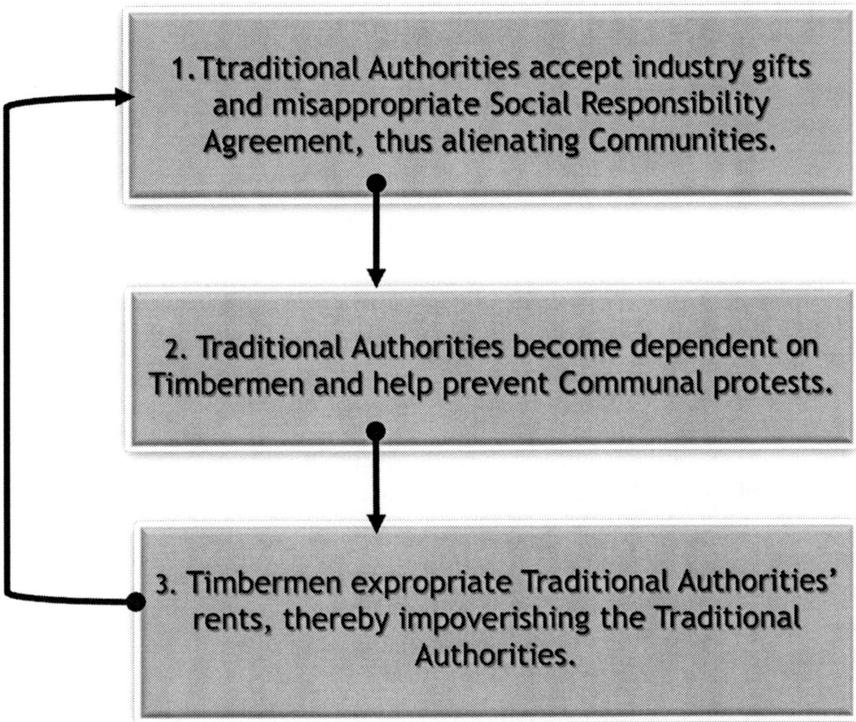

Figure 7.4. Corruption and Power Relations among Traditional Authorities and Timbermen
Source: Adapted from Arthur et al. (2007).

tomarily vested with the responsibility of protecting communal resources. The TAs who operate across the formal and informal sectors, as noted by Arthur et al. (2007), do not have sufficient power to contest state officials' monopoly. However, they possess enough nuisance value to compel the Timbermen to buy them off with "gifts" in cash and/or in kind. The TAs accept such "gifts" or handouts (sometimes they initiate the transaction) because they are often starved of income, which sometimes leads them to misappropriate payments from the community's Social Responsibility Agreements (SRA) with the timber firms, which are critical for the communities. This is because these agreements and related funds help to meet, *inter alia*, some of the communities' basic infrastructural needs. Such misappropriation could thus spark community protests. Accepting such handouts therefore put the TAs permanently under the control of the sponsoring companies or Timbermen. This TA-Timbermen linkage also, inevitably, undermines national policies relating to the timber industry at the local level, as national policies at this level are compromised by such corruption.

The District Forestry Officers Corruption Cycle

As part of the linkage between the key national decision-makers, the politicians, and technocrats, and the "Timbermen," there is often tremendous pressure on the District Forestry Officers (DFOs) by the former group, often referred to as "Accra" (the Seat of the Ghanaian Government), to extend favors to the well-connected Timbermen. Like many government employees, the DFOs are poorly paid and are generally rent-seeking. They are thus willing to accept "gifts" or handouts to make up for their inadequate official incomes. Sometimes these DFOs initiate the transaction via tacit demands, lest they miss an opportunity to make a "quick and easy buck" off the record. They are also protective of the guaranteed government employment and income in a country with a high proportion of unemployed people, which makes them compliant to orders from their superiors. In relation to these, as noted by Arthur et al. (2007, p. 33), they readily comply with orders from "Accra" to facilitate resource access to favored Timbermen, or face sanctions if they do not cooperate. The DFOs are also susceptible to corruption by the Timbermen because oftentimes, their districts do not have operational funds. Consequently, in discharging their official duties, they tend to depend on their own out-of-pocket expenses in fuelling their vehicles or motorbikes, as well as their meals in the field.

As shown in figure 7.5, it is under such circumstances that render the Timbermen not only to be willing to bear their expenses, but also to pay them higher remunerations than the amounts they receive from the Forestry Commission as annual salaries. In the districts, accountability to the state or state oversight is minimal or nil, consequently the chances of the DFOs being exposed and punished are slim, but with their collusion comes the fear of exposure or blackmail by their patrons. They therefore become beholden to the Timbermen, even after they are promoted to senior positions. With the Timbermen in such a powerful position, their payments to the government are significantly lowered to enable them to make more profit, and the reduced state revenues keep the DFOs' salaries down and thus sustain the cycle. As part of this system, some chainsaw operators even duplicate felling licenses and obtain logging concessions though bribery (Quartey, 2010), indicating the formal-informal sectors' inter-segmental movements for gain. These corruption cycles contribute to the marginalization of the forest communities.

MARGINALIZATION OF THE FOREST COMMUNITIES

In 1992, the proprietor of The Makers, one of the small-scale wood processing firms, remarked to the author that "the government has been exploiting

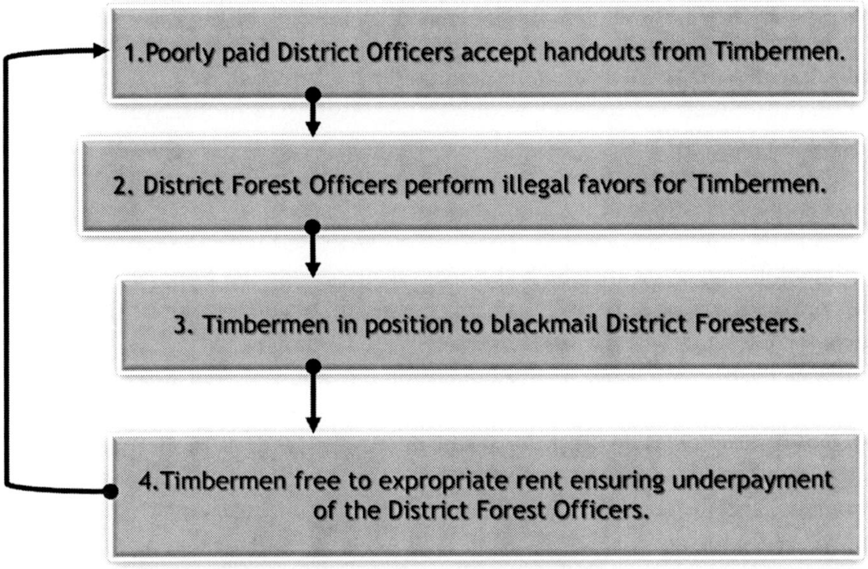

Figure 7.5. Corruption and Power Relations among District Forest Officers and Timbermen
Source: Adapted from Arthur et al. (2007).

and exporting timber as if we are the last generation of Ghanaians"; yet this process of unprecedented exports involved an increased marginalization and neglect of the very peoples whose very livelihood depended upon the source of the timber—the forest communities (Bamfo, 2003). In this vein, the Technical Director of Ghana's Ministry of Lands and Forestry, E. Koffi Smith, succinctly captured the marginalization of the forest communities, noting that, "Over the years an imbalance has crept in, whereby the industrial and national interests have taken precedence over the local community interest" (Koffi Smith, 1996, p. 3). Despite the unprecedented timber exports and associated incomes generated, these "forest-owning and dependent communities receive negligible returns from the destruction of their resources," according to the FERN Report (2006), pointing out that Forest Watch members in Ghana, in fact, believe that community representatives receive less than 5% of their royalty entitlements and none of their entitlement to timber rights fees (FERN, 2006, p. 11). The antecedent to this marginalization is noted in the report that the colonial state, a creation of the European merchant houses, "discouraged traditional subsistence economic activities in a bid to force communities into the cash economy and to cut off their links with their resources" (FERN 2006, p. 19), and establishes continuity of the process by further noting that in the post-colonial era, "The state continues to ensure cheap access to communities' assets on behalf of the corporate sector (especially the transnational corporate sector). The state continues to abuse community rights. The international commodity markets continue to undervalue natural resources and transfer wealth to big businesses in the North" (FERN 2006, p. 20).

Under the adjustment program, the Report pointed out three areas where the "unfair benefits-sharing" with the forest communities is largely reflected (see FERN, 2006, pp. 11–13). The first is associated with the terms of trade in international trade in which the market prices for the exported wood do not reflect the real economic, environmental and social value of the forests destroyed to supply the wood product. While the distorted prices constitute a direct transfer of rents to the buyers or the international consumers, the transfer tends to be directly enjoyed by the formal sector of the industry. The second involves state collection of revenue in the form of timber rights fees, stumpage fees, income tax, and export levy at the expense of the forest communities, and the third involves elite capture of the forestry-related revenues collected by the state.

According to Article 267(6) of Ghana's constitution, 10% of the revenue or royalties accruing from stool lands should be paid to the administrator of stool lands to meet administrative expenses. Out of the remainder, through the traditional authority, the senior-most stool of the traditional political hierarchy, 25% should go to the stool for its maintenance in keeping with its status; 20% to the traditional authority itself; and 55% to the district assembly of the area of authority of which the stool lands are situated. This centralized approach to benefits-sharing is problematic because, as noted in the FERN report (2006, pp.14–15), "none of these institutions accounts to forest-owning communities for royalties they receive; none of these institutions has deployed these resources in development projects that could create long-term economic opportunities that compensate communities for resource destruction."

The elite capture of the royalties, the FERN report notes, is expressed even by the chiefs, who tend to appropriate royalties for their personal or household use. They rationalize this appropriation with a claim that the constitutional stipulation "maintenance of the stool in keeping with its status" refers to that royal perk, and that it is only the royalties allocated to district assemblies that belong to the community. This royal view disregards the fact that the chiefs are mere formal custodians of the stool lands; consequently, compensations to landowners in the form of royalties belong to the community as a whole. This is clearly stated in Ghana's constitution that such "lands are vested in the appropriate stool on behalf of, and in trust for, the subjects . . . in accordance with customary law and usage" (Article 267(1)). As pointed out by the FERN

Report (2006), since stools symbolize the social unit, and not the person who occupies it, the status of the stool can therefore only refer to the wellbeing of the community whom it symbolizes.

Despite the deforestation associated with the expanded timber exports, and the community exclusion from the benefits generated, it is, rather, the members of the community—the struggling farmers and the desperate chainsaw operators (along with bush fires)—who are actually blamed for the deforestation by the Forestry Commission and the forestry industry. In defense of the marginalized forest communities, the FERN Report argues that it:

> is a gross distortion of the truth. While there have, for example, been bush fires (especially in the mid-1980s), these can be purposefully set by the unscrupulous as a tactic to destroy the evidence of over-logging. There is also illegal clearance for farming; but this, too, often occurs in the wake of industrial devastation . . . there is, indeed, a chainsaw problem, but the evidence suggests that chainsaw operators are increasingly funded and answerable to persons close to the established timber (export) companies. (FERN, 2006, p. 9).

Again, despite the fact that the "ownership rights of pre-colonial communities and the applicability of their customary laws to surface land are not in dispute," as noted further in the Report:

> in practice, state management rights have meant expropriation of communities. The only incidences of ownership that these communities still enjoy are the right to receive a share of royalties from forest exploitation and to participate in inventory taking prior to the award of timber utilization contracts. The state takes most important decisions without reference to communities, and discriminates against communities with respect to resource access. Laws dating back to 1927 have criminalized community use of forest resources (FERN, 2006, p. 10).

Even though the Forest and Wildlife Policy (FWP) of 1994 sought to involve all stakeholders for the efficient management of forest resources, the concept of participation was later ignored under the Timber Resources Management Act (TRMA) of 1998. As noted by Odoom (2005, p. 7), despite the fact that the TRMA required farmers' consent to the felling of trees on their farms as well as payment of compensation for felling damage to crops, "the assessment of such compensation payments is not fully transparent and the rights of appeal unclear." Bamfo (2003) sums it up this way: "In the past 50 years, Ghanaian farmers have lost many of their traditional rights to timber trees on farms, and their access and rights to land and trees in reserves have been severely curtailed." Such loss of rights has been exacerbated with the growth of industrial wood processing under the adjustment program.

THE FOREST COMMUNITIES AND THE TURF WARS

The FERN Report (2006, p. 17) showed that the combination of felt deprivation, rights violations, and a sense of betrayal by state institutions led to increasing levels of tension and conflict in forest fringe communities. These tensions, it notes, have in fact exacerbated pre-existing gender, age, and lineage tensions, reflected in increasing domestic violence, chieftaincy disputes, and neglect of the aged, including its extreme community form—accusation of witchcraft. It also showed the crystallization of ethnicity as a class distinction in some places as a result of an increasing tension between earlier settlers or the "indigenes" and later settlers in certain areas. Evidence of the turf wars has become more apparent as the forest communities confront timber companies in an organized manner and demand rent illegally. For example, around the Desiri Reserve in the Nkawie Forest District, many Community Forest Committees—the village-level committees initiated by the Forestry Commission to mobilize support for forest policy goals and to represent community concerns to District Forest Offices, according to the Report, and which are normally chaired by the village headman—set up illegal roadblocks and forcibly demanded "tolls" from passing timber trucks. Also, some of the communities tacitly support raids of company concessions by chainsaw operators—who at least pay rent up front.

The most significant aspect of this turf conflict is that both gangs of illegal loggers and forest-dependent communities are now willing to confront the state over resource access in an increasingly violent manner. As noted in the Report, "There have been repeated reports by District Forest Managers that gangs of illegal chainsaw operators are armed and prepared to resist state interference with their operations" (p. 17). The report further noted that there is no evidence to suggest that this represents community-based "activism" or whether communities are simply responding to and taking advantage of the obvious breakdown in authority. However, it shows that in some places, such as Desiri, the state has actually lost control over the forest reserves and the armed communities have demonstrated a willingness to confront Forestry Commission officials, the police, and the military to protect illegal farming activities. In 2005, for example, the media reported that at Enchi communities engaged in three days of gun battles with Forestry Commission officials and police who tried to drive them out of illegal farms and communities, and kidnapped and seriously injured several police officers in the process (FERN 2006, p. 17).

In order to conceal the state's own weakness, and its repression and abuse of the forest communities in the turf wars, state officials tend to play these incidents down, even though such trends could completely undermine the state, and national integrity, if not checked. In fact, it is the issues relating to state management and decisions, and criminalization of community that link our discussion with the shortage of lumber on the domestic market for use by the community; the informal sector as a context for survival by some members of the community such as the chain saw operators; and the relationship between the formal and informal sector in the domestic and international timber trade. It is also in relation to this link that Sasu (2004, p. 2) argues that as a reflection of the rather contemptuous perception of the informal sector, "not much has been done to promote collaborative resource management due to the reluctance of the state to devolve management authority to local communities," much of which is informal. It is in this context that he notes that since the early twentieth century, succeeding colonial and post-colonial governments—in jostling to control the turfs of the traditional authorities—have attempted to shift control or supervision over disposition of lands from traditional rulers to the state, and that one of the key reasons for the continued centralization is the "perceived notorious traditions of misuse of land by the traditional authorities." Even for the formal sector, the post-adjustment production and export trends entailed that Ghanaian wood-processing firms now operate as subcontractors of convenience for firms overseas *via* an expanded forward linkage, and in the process exporting potential Ghanaian wood-processing jobs overseas.

CONVENIENCE SUBCONTRACTING AND EXPORT OF WOOD-PROCESSING JOBS

As shown in figure 7.2b, an intensified international forward linkage system emerged based on the sustained and dramatic increase in those exports over the adjustment years as shown in table 7.3. Since the exported lumber and furniture parts undergo further processing off-shore, it represented potential jobs that could have been created for Ghanaians, whose natural resources were being exported. This intensified linkage could be said to validate the critics' argument that adjustment would reinforce the Ghanaian wood-exporting firms as sub-contractors for the principal overseas firms (see Dicken and Lloyd, 1990). Ghana's vulnerability in this forward linkage system was demonstrated, for example, by the fact that international buyers even had a tendency to stock up the cheap wood, which later depressed prices further because of diminished demand (Barker et al., 1991, p. 5). Also, the recession in Europe at that time led to a very slow movement of the wood stock in a market already inundated with wood from other African countries (FPIB, 1991). Stocking up on supplies therefore becomes part of the principal firms'

Table 7.3. Volume of Processed Wood Exported,1983–1993 (M^3)

Year	Lumber	Other Processed Wood
1983	44,752	1,942
1984	56,409	2,156
1985	76,406	2,228
1986	83,577	699
1987	161,927	3,440
1988	169,640	5,637
1989	154,033	5,151
1990	200,517	5,595
1991	182,644	8,304
1992	211,427	63,383
1993	224,914	89,755

Sources of Data: Timber Exports Development Board, Ghana; Forest Products Inspection Board, Ghana; Ministry of Lands and Natural Resources, Ghana.

"just-in-case" strategy, rendering the Ghanaian producers very vulnerable to external shock. Thus, as subcontractors, they constitute potential "shock-absorbers" for the European principal and other overseas firms.

The evidence with respect to international linkages, therefore, suggests a reinforcement of the "colonial relationship" between wood processors in Ghana and firms overseas, particularly Europe. Sawyerr's (1990) notion of international capital accommodation, under Ghana's SAP, is clearly demonstrated by the steady and dramatic increase in the volume of lumber and other processed (and unprocessed) wood exported—products that constitute raw materials or inputs for firms overseas. Significantly, the bulk of the increased exports (at least a minimum of 81% per annum until 1990) went to and continue to go to the donor countries, particularly those of the then European Economic Community—now the European Union (Owusu, 1995, 2001c). Ghana's huge quantities of wood were sold in Europe at prices lower than they would have commanded if they were processed further (Barker et al., 1991, p. 5). Thus, with the adoption of the ERP in 1983, a new era was established whereby Ghana became reliable turf in the accumulation process for international capital—a cheap and reliable source of tropical wood, achieved primarily *via* indebtedness from the SECALs and the frequent series of adjustment-related devaluation (see Owusu, 1998b, 2001b).

In sum, since Ghana's adoption and implementation of a structural adjustment program in 1983, the economy witnessed a dramatic turnaround in terms of the volume, content, and direction of its timber trade. The program, as has been shown above, engendered a remarkable expansion in the volume of timber exported; some diversification in terms of content, with lumber replacing logs among the products and further processing such as furniture parts and moldings production taking place along with plywood and veneer exports in terms of content; and finally new markets emerged even though Europe continues to serve as the dominant market. The dramatic increase in the production and exports of processed wood demonstrated "efficiency and growth," a change which is touted as evidence of the "success" of the program in Ghana. The associated changes in the spatial structure of economic activity, including changes in linkage patterns, reflect progress towards the optimum spatial production and activity pattern coincident with SAP-based economic efficiency according to SAP supporters.

To the critics, however, the changes did not reflect the two indicators of spatially integrated development. First, the changes are devoid of dynamic linkages within the formal wood processing industry, and with other sectors in the domestic formal economy (see Darkoh, 1973; Mabogunje, 1981; OAU, 1982). Even the new and intensive formal-informal sectoral linkage within the wood processing sector (absent in the economies of advanced industrialized countries) is largely exploitative. Hence, it hardly qualifies as a substitute for a traditional forward linkage between primary processing and processing downstream within a formal production network (see Fik, 1999, pp. 20–21). In spite of the proponents' argument that the informal activities could generate capital for self-transformation into formal operations (e.g., World Bank, 1989, pp. 135–36), there are no guarantees that such transformation would occur, given their very precarious nature (Ninsin, 1991). Second, a significant decrease in the "colonial relationship" indicating a shift from external over-dependence and asymmetrical linkage relations with international capital (OAU, 1985; UN/ECA, 1986; Sawyerr, 1990) is not apparent. A clear outcome of Ghana's SAP is the dramatic growth of output in a particular sector, which does not fashion real development as expected in a developing economy because the true beneficiary of the changes is international capital (Owusu, 1998a). This outcome is exactly the kind of situation that *The Brandt Report* (Brandt, 1980, 23) referred to as it cautioned that "*one must avoid the persistent confusion of growth with development*" (emphasis mine).

The "success" claimed by the sponsors of Ghana's SAP was, thus, based on a misconception (see Bromley, 1995; Owusu, 1998a), or even a misrepresentation of what development in Sub-Saharan Africa should involve. Interdependence among firms in the advanced Western economies generates various benefits (e.g., cumulative upward causation and its associated agglomeration economies), which are largely internalized because such developed economies (or blocs like the European Union) are intrinsically self-centered. In contrast, the dramatic increase in Ghana's lumber export for further processing off-shore has been achieved at the cost of disrupting the basic intra-sectoral linkages between primary and secondary stages of the formal wood processing industry. That crucial linkage, at the intra-national level, is bypassed to provide huge quantities of lumber cheaply to international capital by exploiting the informal sector while the liberalized economy discouraged formal inter-sectoral linkages. At the international level, the wood-processing firms operated as subcontractors of convenience for the "principal" firms overseas via a strong forward linkage established between them under the program, thereby consolidating the industry's asymmetrical alignment to the interests of metropolitan/donor countries.

Clearly, the SAP-based changes in Ghana's formal wood processing industry did not promote the development of structures for self-sustaining, self-reliant national economic development as expected under the LPA. In fact, Ghana's situation might not be significantly different from other processed tropical wood exporting countries in SSA for at least two reasons. First, at the national level, the other SSA economies have structural characteristics that are similar to Ghana's in the pre-adjustment era (UN/ECA, 1986). Also, like Ghana, political power in most countries is vested in an alliance of social classes with no fundamental interest in the transformation of the local political economy and a break from domination by metropolitan interests (Sawyerr, 1990, p. 10). Second, at the international level, the other SSA economies occupy the same peripheral and subordinate position in the global capitalist economic system as Ghana, with a one-sided external economic orientation, and the associated external dependence (UN/ECA, 1986; Sawyerr, 1990).

Ghana's experience, based on its formal wood processing industry, suggests that SAPs stifle spatial integration at the intra-national and international levels, and sustain dependence on foreign capital and markets. This is perhaps because, simply put, the IMF, the World Bank, and the lender or donor countries and agencies are neither charitable organizations, nor are they neutral in their activities. Their own interests as international financiers, or their need for cheap and reliable sources of industrial raw materials, necessarily conflict with the emergence of structures for self-sustaining, self-reliant national economic development in peripheral countries. Consequently, African development within the broader global economy should be designed primarily by the Africans themselves as reflected in the LPA and the African Alternative Framework to Structural Adjustment Programmes for Socio-Economic Recovery and Transformation (AAF-SAP). A more pertinent development approach to keep African economies from further decline must address their respective adjustment and structural transformation problems simultaneously (see UN/ECA, 1986, p. ii).

Like the goals of the AAF-SAP, this approach must be "holistic in nature" so that "the macro-economic framework, the policy directions and measures, and the implementation strategies take into account the dynamic relationships existing among all major elements related to adjustment with transformation. Thus, the dichotomy between structural adjustment and long-term development is eliminated" (UN/ECA, 1986, pp. ii–iii). While the peculiar characteristics of each individual country should be taken into account, great emphasis must also be put on the full mobilization and efficient utilization of domestic resources, including government assistance to the informal sector (UN/ECA, 1986). This approach, which is more consistent with the LPA, would contribute significantly to the development of spatially integrated economies in the region, and link them to the global economy in a mutually beneficial way. Unlike the Ghanaian case, where the adjustment program was adopted as a metonym, this approach requires a genuine commitment on the part of African governments to real structural transformation of their respective economies.

In addition, Ghana's quest to sustain and in fact, increase its share of the international trade in timber, beyond the associated unprecedented deforestation, further marginalized the forest communities, thus jeopardizing their very livelihoods. In terms of spatial integration, the quest also created a serious domestic lumber shortage, which necessarily pushed the informal sector, evidenced by increased chainsaw operations, to step in to meet the domestic demand, and a constricted traditional local forward linkage between sawmills and downstream processors in the formal sector. At the international level, the program turned the Ghanaian formal wood-processing firms into subcontractors of convenience for firms overseas via an expanded forward linkage, thus exporting potential Ghanaian jobs overseas. Policy-wise, one cannot look at a country's international trade without due regard to the material source of the trade and its socio-economic framework, namely the forest, its peoples, and their livelihoods. In this respect, it can be said that at the root of all these new trends in Ghana's international timber trade is a basic disconnect between the actors, namely the policy makers who operate largely in the formal sector which is driven by the dynamics of capitalism on one hand, and the stakeholders, mainly the poor including the forest communities and others, who largely straddle the formal and informal sectors for purposes of survival. It is therefore very important for the actors in policy formulation and implementation, to remember that traditionally (for centuries before colonialism), these stakeholders have depended on forest resources for their livelihoods—their economic production, social organization, religious beliefs, and identity have all depended on this relationship. Consequently, the formal and regulated exploitation of the forests by the government, and the exclusion of their communities from access to these forests are very disconcerting. A question by an elderly man, Opanin Dwomo of *Sagyimase*, a forest community in the East Akyem area, is quite instructive in this regard. In his frustration with the formal sector—the government and the timber industry, which essentially were expanding their turfs at the expense of the forest communities rooted in the informal sector—asked a researcher: "Why do we need permits to enter the forest reserves to fell trees for the construction of homes? And why does the government not do the same to our brothers who live by the sea? We do not understand this" (Sarfo-Mensah, 2005, p. 16).

In this regard, for policy makers, there should be an on-going recognition that a healthy symbiotic relationship between the formal and informal wood sectors in the economy as well as the forest communities is critical for the sustainability of the timber industry, as a whole, and the associated domestic and international trade. This requires a dedicated effort for cooperation with the forest communities to combat deforestation, which is in the mutual interest of the actors and stakeholder; the forest communities must be given some formal access to timber for home construction and other basic needs; and the rural folks should be integrated into wood-processing activities to provide the youth, in particular, with jobs. Finally, given the widening markets and increasing volumes of processed wood sales, a significant proportion of revenue should be expressly set aside to help forest communities exclusively in the form of social and economic infrastructure, and scholarships or financial aid for education for the youth. Such arrangements could be supported by agreements with major consuming countries in Europe, emerging ones like China (which strategically need African timber), the United States, and the nouveaux riches ones like Saudi Arabia, Kuwait, and the United Arab Emirates.

Another measure whereby the government sought to control its turf was by becoming a signatory of *The Year 2000 Objective*, as was noted in chapter 6. This was initially signed by the PNDC government and continued by the NPP administration. In compliance with the declaration, the government sought to reduce pressure on the natural forest through harvest limits, the use of an ideal mix of species, and efficient use through value adding processing. One of the government's strategies for offsetting pressure on the natural forest involved decreasing the annual allowable cut (AAC) and encouraging diversification of species used. In addition it also introduced initiatives geared towards improving woodworking skills which improve utilization efficiency. Furthermore, it discouraged the export of air-dried products in favor of kiln-dried products, as a measure to promote in-state value-added processing further, with some preferred species exported as air-dried lumber facing levies ranging from 10 to 30% FOB, depending on rareness of species (MLF, 1996b). This meant that the government was encroaching on the turf of the Ghana Timber Millers' Organization (GTMO), which criticized the government, complaining that the industry was on the verge of collapse because of the extensive government interventions and levies (GTMO 2000). This was in spite of the fact that on the international arena, Ghana was listed in 2000 as one of the top six countries—"The Super Six"—which appeared "to be managing some of their forests sustainably at the forest management unit level to achieve the *Year 2000 Objective*" (ITTO, 2000; Poore and Chiew, 2000).

In connection with the *Year 2000 Objective*, was Ghana's pursuit of a Validation of Legal Timber Programme aimed at improving the monitoring of tree harvesting and capturing every single tree in a database for verification; a Timber Utilization Contract system for social responsibility of concessionaires; together with other social responsibility arrangements, as well as its commitment to a voluntary partnership agreement with the EU to improve its trade and production of legal timber are all laudable. However, without addressing the partnership with the informal sector and the constituent forest communities, success in these areas will surely be very difficult, if not impossible to achieve. Fostering such a healthy symbiotic relationship requires a new, pragmatic, and sympathetic look, by African governments and policy makers, at the true nature and dynamics of the informal sector and the forest communities. In fact, the difficulties that are encountered by the government in banning the operations of chainsaw operators in Ghana is testament to this need. This way, policies formulated and implemented would be based on partnership, a necessary ingredient for the sustainability of the forest sector as a whole.

In the next chapter, we focus on the dynamics of adjustment-related debt and dependence relative to the timber industry and the Ghanaian economy in general, and the implications for comparable African countries. We will explore how the adjustment process enabled international capital to create, sustain, and control its turfs by ensuring that the turfs become dependent through the encouragement of debt both on the domestic and the international levels.

Chapter Eight

Debt, Dependence, Turf Control, and Ghana's Economy

In chapter 7, we focused on the new patterns of trade in terms of the sourcing, volume, content, and direction of its timber trade, and the associated turfs that emerged as a result of the program. We showed that generally, the program engendered a dramatic up-turn and expansion in Ghana's timber exports, especially lumber, to the traditionally dominant European market, with new markets, including the Middle East, Asia/Far East, North America, and some West African countries also steadily becoming important. We discussed the factors or determinants that engendered the upswing in production and exports and the associated new trends. These new trends however were shown to have had largely negative effects not only on the environment as we saw in chapter 6, but also on the traditional interdependent relationships between the formal and informal sectors of the wood industry and on the very livelihood of the forest communities and forest management. At the domestic level, it was shown that Ghana's quest to sustain its wood exports has constricted the traditional forward linkage between sawmills and downstream processors and created an acute lumber shortage on the domestic market, thus generating what is largely perceived to be illicit activities by chainsaw operators to meet the domestic demand, and further marginalizing the forest communities. At the international level, it was shown that Ghanaian wood-processing firms now operate as subcontractors of convenience for firms overseas *via* an expanded forward linkage, and in the process exporting potential Ghanaian wood-processing jobs overseas. These new trends and their implications were tied to adjustment-related loans and the associated economic dependency it engendered.

ADJUSTMENT, DEBT, AND THE FORESTRY SECTOR

For international capital, Ghana's "forestry sector offered the greatest immediate potential for growth and foreign exchange earnings" (World Bank, 1984, p. xx). Government policy under the ERP with respect to the roots of the wood-processing industry, the forestry sector, was to ensure improvement in that sector's deteriorating facilities, proper husbandry, and efficient utilization of the country's forest resources (Forestry Commission, 1988, p. iv). In connection with that, the government aimed at "achieving a quick increase in production of the private companies through providing them with spare parts, equipment and materials," supporting the rehabilitation of selected mills, whilst re-organizing state-owned ones (Govt. of Ghana, 1984, p. 11). The government also moved to alleviate the immediate physical constraints in the form of production equipment shortages, transport equipment, and spare parts with the assistance of SECALs from the International Development Association to fund the export rehabilitation project (the real "ERP"). There were other SECALs from such international organizations as the Overseas Development Administration (ODA) through which the British government provided the PNDC government with some £8.7 million in 1984 and 1985, as a grant, tied to the purchase of timber sector equipment from U.K. manufacturers. Some of the SECAL facilities were offered with a food aid program for the employees of the beneficiary companies to make the loans more attractive, thus exploiting the "food crisis" situation of the time.

DEBT AND DEPENDENCE IN THE ADJUSTMENT ERA

In general, in terms of the mutual needs of the Ghanaian government and international capital, the program could be considered as successful, as clearly demonstrated by Owusu (1998b). The rehabilitation of the forestry sector led to a dramatic increase in the production and export of wood and wood products (see TEDB, 1991; FPIB, 1991; Engeberg-Pedersen et al., 1996). In meeting the needs of the PNDC government of Ghana, therefore,

> the adjustment program and the contributions of the forestry sector have been a convenient means of accessing international loans by actively servicing the country's debts with foreign exchange generated through the export drive. In 1986 the share of official debt (i.e., government and government guaranteed debt) in total long-term debt was 85%. In 1992, it had risen to 93% (World Bank 1993) indicating increasing indebtedness as result of easier access to international loans. In accommodating international capital via the program, the PNDC government was able to stay in power by defusing the political crisis it

> faced prior to adjustment. It has even been able to transform itself from a military government into a "democratically elected" one—the National Democratic Congress (NDC). It enjoys "political stability" with support from the West, and has become a protégé of the sponsors, making Ghana a "star pupil of adjustment" in the eyes of the World Bank and the IMF (Owusu, 1998b, p. 433).

For the external sponsors or international capital, Owusu further noted that

> the adjustment program and the contributions of the forestry sector have been a convenient means of generating cheap raw materials—tropical wood and wood products—in abundance, for their home industries. The loans advanced to Ghana by the sponsors, and the interests on them are serviced consistently. . . . Furthermore, the wood processing sector has served as a convenient market for products like machinery and equipment from some of the donor countries, and the administration of the SECALs was a convenient way of sustaining, and in fact reinforcing, the market (Owusu, 1998b, p. 433).

Let us now examine closer the generation of debt and the accommodation of international capital in the form of reinforced dependence in the Ghanaian economy and skewed benefits to international capital.

INCREASED GROWTH IN DEBT

Based on the World Bank's own figures, table 8.1 shows the level of Ghana's outstanding external public debt and the debt service payments made prior to adjustment in 1983.

The table shows that by the beginning of 1983, the disbursed funds had grown from less than a billion dollars in 1980 to US $1.12 billion, whilst the undisbursed funds continued to hover around US $1.5 billion. At the beginning of 1983 the service payments had reached their highest, and meeting the debt servicing obligations was in jeopardy. Table 8.2 shows the rapid growth

Table 8.1. Ghana: Pre-Adjustment External Public Debt and Service Payments (1980–1983) in Thousands of US Dollars

	Outstanding Debt at Beginning of Period		*Service Payments*		
Year	*Disbursed Only*	*Including Undisbursed*	*Principal*	*Interest*	*Total*
1980	983,899	1,403,638	65,553	28,382	93,935
1981	1,100,426	1,534,974	28,255	24,966	53,221
1982	1,114,020	1,503,523	34,399	27,778	62,177
1983	1,120,638	1,499,479	61,041	38,705	99,746

Source: World Bank (1985,33).

Table 8.2. Growth of Ghana's External Public Debt and Service Interest Payments: 1982–1991 (US $ millions)

	External Public Debt		
Year	*Disbursed Only*	*Including Undisbursed*	*Service Interest*
1982[a]	1,114.020	1,503.523	27.778
1983[a]	1,120.638	1,499.479	38.778
1984[b]	1,152.058	1,592.809	29.238
1985[b]	1,299.464	2,103.879	30.458
1986[b]	1,721.722	2,561.811	42.624
1987[b]	2,255.602	3,348.105	53.782
1988[b]	2,192.372	3,273.738	69.166
1989[b]	2,335.904	3,597.185	59.499
1990[bc]	2,531.508	3,393.848	67.277
1991[bc]	2,756.985	3,270.553	70.637

Sources: World Bank (1985, 3; 1991, 99).
Notes: [a] Outstanding at the beginning of period.
[b] Outstanding at the end of period.
[c] Projected based on debt outstanding including Undisbursed as of Dec. 31.

of the external debt and service interest payments after the adoption of the ERP in 1983.

It now became much easier to access foreign loans as shown by the growth from the US $1.5 billion mark to US $2.76 billion (disbursed) and US $3.27 billion (including undisbursed funds), and the service interest payments rising rapidly from the US $38.7 million of 1983 to US $70.6 million in a span of eight years. According to Bank of Ghana (ISSER 1995, 63), interest payment in 1990 was US $111.2 million, rose to US $119.2 million in 1991, and was at US $114.7 million in 1993. The corresponding total payments (including the principal amounts) were US $375.3 million, US $335.0 million, and US $464.5 million respectively. Again, the Bank of Ghana figures show a total debt of US $3.79 billion at the end of 1991, which had grown rapidly to US $4.85 billion by the end of 1994 (ISSER 1995, 62). Both the World Bank figures and Bank of Ghana's figures are, at least, in agreement in one area: Ghana has accumulated substantial amounts of external public debt since the adoption of the adjustment program, and the associated debt servicing payments increased rapidly thereafter.

SECALS AND THE ACCOMMODATION OF INTERNATIONAL CAPITAL

With the infusion of SECALs into the forestry sector, production and exports in the formal wood processing industry were tremendously boosted, as was anticipated by the proponents, in consistency with their "efficiency" argument. Due to the rehabilitation of the forestry sector, the production of logs increased dramatically, registering a change of 290% over the short period of 1983 through 1990. As a proportion of total production, log exports increased from 13.8% in 1983 to 15.3% in 1990. Not much value was, thus, added to the wood exports; this was in spite of a government policy of curtailing log exports in favor of processed-wood. There was an increase of 354% in the volume of logs exported. Primary-stage processing (lumber production for export), involving minimal value added for the Ghanaian economy, continued to dominate in the wood processing industry, registering an increase of 325% in exports over the period. Relative to total output, the volume of lumber exports increased steadily from 25.9% in 1983 to 37% in 1990, thus steadily rising towards the annual minimum of 51% that characterized the 1960s. Processing further downstream (e.g., furniture and fixtures, plywood, and veneer), however, continued to be small in terms of size and growth with the volume of exports increasing by only 148% (Owusu, 1994). It must be noted that it was the high quality wood that was exported. Local needs were largely satisfied by second- or lower-grade wood because local consumers could not afford the high quality wood at the international market prices.

Major reasons for the expansion of wood exports include the fact that the SECALs were advanced primarily to rehabilitate the industry to produce raw materials for export. This was particularly so, with the infusion of as much as 70% of the rehabilitation capital into primary-stage processing or sawmilling, the output of which serve as inputs for overseas firms. In addition to these, the incentives under the investment code (PNDC Law 116, 1985), such as the provisions for profit retention abroad, the huge tax exemptions based on volume of output exported (see UNIDO, 1991, p.52), together with the flexible exchange rate system, *inter alia,* encouraged firms to focus primarily on exports. Furthermore, under tremendous pressure, the SECAL-beneficiary firms had no option but to increase output as much as possible, in order to generate enough money, in the face of the series of devaluation and rising interest rates, to service their debts (see Owusu, 1998b, 2001a).

As was expected by the critics, the exploitative role of international capital was, thus, reinforced during the period, for raw materials continued to be exploited in a more intensive manner *in situ* as indicated by the steady and dramatic increase in the volume of processed and unprocessed wood exported. Indebtedness particularly from the SECALs and the frequent devaluation of the Ghanaian currency, which cheapened the exports, compelled firms to produce and export more processed-wood in order to survive. As was the case prior to the ERP, the bulk of the increased exports went to the donor countries, particularly the EEC countries, which received a minimum of 81% per annum. It should be borne in mind that lumber and furniture-parts, which constituted the bulk of the exports, are essentially raw materials for firms in the metropolitan or donor countries, and despite the policy of phasing out the export of unprocessed wood, the volume of logs exported increased. Significantly, these raw materials became so cheap as a result of the frequent and massive devaluation of the currency that, often, resale took place on the world market (Barker et al., 1991, p. 5). There was even a tendency among some overseas buyers to stock up on the cheap wood, which further depressed prices because of diminished demand relative to supply. The recession in Europe also led to a very slow movement of the wood on top of the inundation of the international market with wood from other African countries (FPIB, 1991). Stocking up supplies was thus part of the principal firms' "just-in-case" strategy, which rendered the Ghanaian producers very vulnerable—as dependents on external markets—to external shock. In other words, as subcontractors, they constituted potential "shock-absorbers" for the European principal firms (Owusu, 1998a).

The deepened dependence of the formal wood processing sector on external markets entailed the disruption of the traditional and crucial forward linkage between the sawmills and processors downstream like the production of furniture and fixtures as discussed in chapter 7. Such forward linkages, it was noted, are crucial for integrated national development (Mabogunje, 1981). It was, however, that crucial linkage that was substituted—under the adjustment program—with a new and intense external forward linkage, as part of the external market dependence in which Ghana's sawmills served as subcontractors, and produced inputs for principal firms abroad. Secondary-stage wood processing, also, became principally linked and dependent externally since the main export items were furniture-parts, which were sold and assembled outside Ghana (Owusu, 1995, 2001c).

It was to facilitate the exploitation of the raw materials *in situ* for profits that the Ghana Timber Marketing Board was replaced by the twin bodies—Timber Exports Development Board and Forest Products Inspection Bureau—whose setup and organizational structures were recommended by Silviconsult, a consulting firm appointed by the World Bank. This may have influenced Silviconsult's recommendation to be consistent with the economic interests of the World Bank. In any event, the operations of these twin principal organizational bodies of the industry streamlined timber export processes to make them easier and more efficient. Their operations were enhanced by the activities of other related bodies as the Ghana Investments Center (GIC), which promoted the very liberal investment code; and the Ghana Exports Promotion Council (GEPC), which *inter alia*, promoted labor training for wood-processing for export.

The rehabilitation aspect of the program was also very heavily biased towards the restoration of basic infrastructure related to the export of raw materials as argued by the critics. While it is true that the restoration of such infrastructure is indispensable to economic recovery in any event, the ports, the railroad system, the trunk roads from the forests to the ports, and telecommunication system attracted disproportionate amounts of the new public investment. For example, under the PNDC government's Three Year Investment Program (1984–1986) for rehabilitation involving US $ 3,920 million, US $1,200 million (30.6%) and US $ 1,250 million (31.9%) were allocated to physical infrastructure, and fuel and power respectively (World Bank, 1984, p.50). Also, at the Donors' Conference in November 1983, for instance, it was decided that two-thirds of the government's foreign exchange expenditure over 1984–1985 was to be spent on rehabilitating the railroad network, roads, and the ports at Tema and Takoradi, and as much as US $13 million was allocated for the rehabilitation and development of posts and the telecommunications (Government of Ghana 1984, 13). It is in that respect that Sawyerr (1990, p. 30) pointed out such rebuilding of infrastructure tends to reinforce the commitment to, and dependence on, the export of primary products. For one thing, the nature of investments in such infrastructure and the procurement methods adopted in connection therewith involved the importation of machinery and other finished goods, and contractors and consultants from the donor industrialized countries. For another, while the assured supply of such products certainly benefited the donor countries, the benefits to Ghana of continued and increased production of primary commodities for export are less clear in the long run, but even in the short run, the already low export prices give cause to worry.

The administration of the SECALs was a convenient way of sustaining and, in fact, reinforcing the market. For Britain, in particular, it was a very convenient way of using the ODA and credit from the Export Credit Guarantee Department to bail out the then ailing British Leyland Corporation. This was especially successful with the system of tied-purchases and credit line system as pointed out by Hancock (1989). In addition to selling the products, some of which were inappropriate for conditions in Ghana, to the sector, spare parts had to be subsequently obtained from the same sources for their maintenance, as well as other joint/complementary-demand products. In certain cases, technicians from the seller-companies had to be brought in at the expense of the buyers to either install or repair the machinery or equipment. In quite a number of cases, the equipment purchased broke down quite frequently. The tied-purchase and credit-line systems thus propagated an international "junk-yard" syndrome, whereby inappropriate and substandard machinery and equipment or technology are dumped in developing countries for profit, and reinforced the dependency of the industry on foreign technology and inputs (Owusu, 1995). The SECALs also reinforced the pattern of exporting the processed wood to the donor countries, i.e., dependence on external markets. The wood sector and the wood processing industry thus continued to be aligned to the interests of metropolitan/donor countries such that the asymmetrical colonial relations were sustained and reproduced.

Linked to the generation of debt and the dramatic expansion in exports was the need to satisfy international capital by meeting the debt service obligations with respect to both the outstanding and new debts. As noted by Owusu (1998a, 1998b), at the end of the 1985–88 SECAL program, the foreign exchange generated by the industry was enough to absorb the nation's annual debt interest payment as wanted by international capital. For 1991, in addition to the ability of the foreign exchange generated by the sector to absorb the interest, it was enough to contribute US $58.363 million of that year's principal of US $122.714 million. By June 1990, eight years of the ERP, Ghana had been able to pay up all its remaining recorded arrears in foreign debt, thus restoring normal economic relations with the creditors. In terms of foreign exchange generation, therefore, the sector became a net contributor to the payment of foreign debts, a departure from the pre-ERP pattern. In spite of this, the foreign exchange spent on acquisition of imported inputs continued to be high. This was because the adjustment program also intensified the dependence on foreign inputs, namely technology, machinery and equipment, as well as non-wood material inputs, draining the country of the remnants of the foreign exchange (forex) generated. With the trade liberalization policy and the easy access to foreign exchange through the forex bureaux system, it became rather easy for merchants engaged in the importation of the non-wood inputs to flood the domestic market with such items as nails, varnishes, and glues—items which could be locally produced. The forex bureaux are authorized private dealerships in foreign exchange transactions, which replaced the cumbersome pre-adjustment exchange control system, as part of the liberalization process. Consequently, the use of such a system enabled many international financial transactions, such as forex for imported inputs, to be carried out with ease.

As noted by Sawyerr (1990, p. 33), such inputs have a ready market and therefore move quickly and yield high profits. The merchants were, therefore, able to mobilize large amounts of the local currency, the *cedi*, to outbid the local manufacturers who wanted to import capital goods and equipment. Specifically, their competition, apart from bidding up the price of foreign exchange, tended to push the price of foreign exchange beyond the reach of importers of capital goods and other items essential to the industrial recovery of the economy, and the social wellbeing of Ghanaians. The competition, therefore, biased the pattern of imports away from essential items. Also, it tended to flood the market with finished consumer goods, thereby squeezing out local manufacturers, and contributed to the collapse of local industry, whilst it deepened the dependence of the economy upon imports (see Sawyerr, 1990, p. 33).

On top of these, old and used machinery and equipment continued to be imported as before, thus providing opportunities for some overseas firms to dispose of their obsolete machinery and equipment in the sector at a profit, whilst installing new ones based on the latest technology in their own outfits. In fact, the SECALs provided opportunities for both foreign and local firms to engage in seriously corrupt practices that hurt the Ghanaian economy. As noted by the then PNDC government's Secretary for Finance and Economic Planning and the chief architect of the ERP, Dr. Kwesi Botchwey, with regard to "Tied-Aid,"

> even though an important advantage or effect of a shift from controlled to market based system, apart from its benign effect on economic aid incentives through the better rewarding of efficiency, is that it removes a major source of corruption and patronage. Public officials engaged in the allocation of foreign exchange administratively have a strong incentive to demand rewards when they know how much of a premium the beneficiary gets from the exercise of their administrative power (Ghana Today, May, 1993, p. 26).

According to the surveyed firms in our study, the government and the donor agencies were so eager to disburse the funds that companies were inadequately screened and hastily approved for the loans. Mr. Yirenkyi of Bibiani Lumber and Logging Company, for instance, pointed out that the term "food aid" was sometimes used as a bait to lure firms into accepting the loans at that time when food shortages were frequent. The circumstances were such that it was very easy to misrepresent firms in order to get the SECALs by presenting "cooked" feasibility reports, whilst the evaluation officers could be easily bribed or corrupted, as noted by Dr. Kwesi Botchwey. Also, because the suppliers' credits that became available through the SECALs were a direct arrangement between the suppliers overseas and producers in Ghana, disbursement could easily be effected to benefit both parties in ways that benefited them at the expense of the Ghanaian economy. The program, therefore, became a means for stashing part of the credit in private accounts or diverting huge sums of money into private bank accounts or pockets. For example, as was reported in the *Financial Times* of February 8, 1989, in an article captioned "Forests Pay as Ghana Loses Out," the corruption involved both local and foreign firms, including a consulting firm hired by the World Bank (see also Hayter, 1989, p. 120). According to the paper,

> A consulting firm, Silviconsult, was appointed by the World Bank. It directed that a Timber Export Development Board should be created to oversee a network of accredited agents to promote Ghanaian timber abroad and obtain the

> best prices. Many such agents have been shown as consorting with the Ghanaian timber merchants by accepting orders from sister front-companies at low prices. The same sister companies are accused of over-pricing machinery and equipment, paid for by international loans, while claiming to be foreign manufacturers and suppliers.

The corporate interviews conducted as part of our study involved top officials from such companies as Maxwell Owusu Timbers Ltd., Pan Timbers Ltd., and Bibiani Lumber and Logging Company. All these companies were also beneficiaries of the SECAL program. The top management officials of these companies were of the opinion that the loans advanced by the Overseas Development Administration (ODA) of the United Kingdom to rehabilitate the timber sector was one of the British government's strategies for revitalizing British Leyland Corporation, which was then on the verge of collapse. Hence, the ODA's loans were tied to the purchase of equipment and spares from the ailing corporation to its loans (see Hancock 1989).

Another article titled "Ghana Builds on Timber Revival" in the June 4, 1987 issue of the *Financial Times* confirms the nature of the ODA contributions in terms of tied-purchases benefiting the British (see also Hayter, 1989, p. 119). According to the article,

> British supplier Forester, Wadkin, Stenner and Wells won a contract worth nearly $1m for sawmill and other timber processing equipment. . . Another, even bigger, contract is close to conclusion. . . . Britain's Marktrace Projects, 51% owned by a Babcock International subsidiary, last March began work on a five-year management and procurement contract for rehabilitation of the state-owned African Timber and Plywood Company, which has a 6,000 sq. km concession in Samreboi. The $30m project is financed by the Bank of Scotland and Badea (the Arab Bank for Economic Development in Africa).

In general therefore, a greater part of the foreign exchange generated ended up as payments to the donor countries. The real costs of importing capital equipment, machinery, and spares, as well as intermediate inputs increased as part of Ghana's external dependency, thus siphoning off a significant part of the foreign exchange that had been generated by the industry. Again, this was especially facilitated by the investment code, which made it easy to import such machinery and equipment duty-free. The code also made it easy for the foreign technical experts brought in, either independently by firms, or as part of the SECAL to draw their benefits from the foreign exchange generated or have them transferred through the services of the foreign exchange (forex) bureaux system.

The evidence in the foregoing discussion shows that essentially, the program is rather self-serving for international capital. It is apparent from the discussion that, rather than fashioning structures for self-sustaining, self-reliant national economic development, Ghana's SAP—the Economic Recovery Program (ERP)—by its very nature, had an inherent bias towards the reinforcement of the colonial patterns of economic control and external dependence, propagated through such mechanisms as SECALs. Ghana's "comparative advantage," with respect to formal wood processing, the evidence suggests, worked to its own disadvantage, and to the advantage of international capital (Amin, 1990; Barker et al., 1986). This should not be particularly surprising because, as was pointed out earlier, the IMF, the World Bank, and the donor countries or agencies are neither "charitable organizations" nor are they neutral in their activities. These principal actors, namely the international financial institutions, donor countries, and agencies need to ensure the repayment of the loans they have advanced, and that profits are made on their investments. In the case of the sponsoring countries, they also need to ensure that raw materials for their home industries become cheaply and easily available; and that markets for their products are maintained or expanded (Sawyerr 1990; Owusu, 1995, 1998a). As pointed out by Sawyerr (1990, p. 28), "multilateral financial institutions, as agents of 'multinational imperialism,' work to ensure that the economies of the Third World are conditioned to 'accommodate'" international capital. SAPs, thus, constitute a strategy for the realization of those objectives.

Furthermore, given the undeveloped structural characteristics of the developing economies of SSA, unfettered market systems have a strong tendency of amplifying the asymmetrical economic and power relationships between them and the advanced economies, who promote adjustment as a development strategy for developing economies (see Myrdal, 1957; Mkandawire, 1988). This asymmetrical relationship is exacerbated by the laissez-faire principle underlying the program, and the retreat of government activism, which would otherwise provide a countervailing force. Consequently, the relationship is manifested in a process of cumulative downward causation (Myrdal, 1957), whereby the rich get richer at the expense of the poor. For the poor SSA economies, the process is shown to have occurred at both the international and intra-national levels. At the international level, the impact is in the form of increased national indebtedness and external dependence. At the national level, on the other hand, the impact has been in the form of economic distress among firms operating on externally sourced loans, such as SECALs, and what may be termed as economic cannibalism. Economic cannibalism involves the squeezing and ultimate gobbling up of small import substitution industries that produce inputs for local industries, either through the imports by rich merchants under the liberalized market system or by larger firms flooding the market with much cheaper goods.

In this chapter, we have shown that if Ghana is really identified as a "model" adjusting country by the external sponsors of SAPs, then certainly, orthodox adjustment does not suggest itself as the best remedy for the economic problems in SSA, as claimed by the proponents. The evidence from our study rather suggests that, as a development policy for SSA economies, SAPs, like the classical "Trojan horse," could lure adjusting economies, through such mechanisms as SECALs, into accumulating substantial external debts. Ghana's case suggests that adjustment may provide a means for international capital to reap huge profits; secure assured and increased supplies of cheap raw materials; and establish dependable markets of convenience. It also lends credence to the view that adjustment, as a trap, constitutes an excellent basis for international capital for non-political, non-military manipulation of political and economic decision-making in the adjusting countries (Sawyerr, 1990, p. 28). It is hard to envision that, under adjustment, the experience of other African economies would be any different, given their general dependence on foreign loans, export markets, and earnings, as well as the dependence of most of their industries on foreign capital and inputs. Indeed, given the objective position of African economies as a subordinate component within the global capitalist economy, and the tendency among African political leaders to adopt neo-liberal measures or SAPs as a metonym, it appears that any successfully "adjusted" African economy would continue to be even more indebted, and externally dependent and dominated. Of course, that is assuming that such "success" could be attained. In the next chapter, we briefly summarize each of the chapters, make some concluding remarks based on our discussions, and offer some related lessons for the twenty-first century.

Chapter Nine

Conclusions and Lessons for Twenty-First Century Africa

In the preceding chapters, we discussed the structural and organizational changes that have occurred in Ghana regarding the impact of colonial and post-colonial national and international development policies (*e.g.,* modernization and neo-liberal structural adjustment) on industrialization and economic development in Ghana in particular and Sub-Saharan Africa in general. We focused on the role played by a Tropical *Timber* as a natural resource that is supposed to be renewable and sustainable; its exploitation and processing, and the associated issues in the various scales of *Turfs* or operational milieux in terms of domestic and international capital in the developing world of Africa; and the associated processes, interactions, and effects, namely its domestic and international *Trade* patterns over time—the colonial and neoliberal eras. In relation to these, the discussion was couched in the important but often unrecognized or neglected role that geography plays in relation to the unequal relationship between the advanced and developing economies of the world, and how that relationship affects economic development and trade behavior of developing economies. In examining these relationships, Ghana's timber industry was used as an integrative and representative economic sector for our discussion because it typifies the structural and organizational—*Turfs and Trade*—issues like the implications of the domestic and international timber trade for local livelihoods and sustainable forest management, associated with underdevelopment.

In essence, therefore, the discussion focused on the issue of structural transformation—the positive changes or progress made in the region's economies, resulting from domestic and international policies aimed at correcting the systemic economic and social imbalances that tend to obstruct development—in Ghana, and by extension, Sub-Saharan Africa. The core argument made in that respect was rooted in the perspective that the root of Africa's persistent problem of underdevelopment is the very structure of its political economy.

This chapter concludes the discussion by tying together the issues of development, particularly spatial integration in Africa with respect to the tropical timber trade, formal and informal operational turf creation, and their contestations by various socio-economic interests, from the colonial period to the neo-liberal era. It also identifies the lessons that can be gleaned from them for the benefit of students and practitioners of Ghana's development and other comparable African countries in the twenty-first century.

The lessons distilled from the discussions in this book are very important, especially since the twentieth century was seen as "a century of lost opportunities for Africa" because despite all the hopes after emerging from colonialism, it had not attained any appreciable level of development by the end of the twentieth century as a result of a combination of some endogenous and exogenous factors (Konadu-Agyemang et. al., 2006), aspects of which have been discussed in this book. The lessons here are meant for African leaderships and other key actors—domestic and international—to *learn* from the mistakes made in the twentieth century as much as possible so as to avoid *repeating* them, in order to make use of the opportunities right at the beginning of the twenty-first century for a better future. In this vein, the observation of George Santayana, the Spanish American philosopher, is instructive. To paraphrase him, African leaders, policy makers, and stakeholders need to take cognizance of his famous advice that "Those who fail to learn from the mistakes of their predecessors are destined to repeat them" (Life of Reason, Vol. I), if Africa is to make any headway in the twenty-first century. This is because, as T.S. Eliot (1935) points out in his anthology, *Four Quartets*, "Time present and time past; Are both perhaps present in time future; And time future contained in time past." Going forward, there are a few conclusions and lessons that could be distilled from the discussions and conclusions from the preceding chapters.

Chapter 1, Africa, Turfs, and Trade: Theoretical and Historical Framework, provided the theoretical framework and various concepts and processes critical to the explanation of the issues which are raised in the book. This included the related conceptualizations of space and such relevant processes as colonialism. It also identified spatial integration as a critical goal of structural transformation in Sub-Saharan Africa, given the fact that all advanced countries are largely functionally linked internally in terms of their economic activities. A proper conceptualization of the global capitalist space and its constituent African spaces by key policy makers and African development-related actors is important because these spaces must be interpreted as structures created by society, which define the issues which they must be concerned with. These issues include changes in both the administrative and economic organization of space resulting from the penetration of forces from beyond the continent and their interactions with internal structures, as noted by Mabogunje (1981). This is underscored by Bell (1986, p. 16), who points out that the impact of spatial organization on the lives of people—the African people, in our case—must be of central concern to key actors because it is critical to identify the changes in social and spatial access to the means of production and social provision, along with the patterns of social and spatial inequalities which have emerged. Further, she notes that it is equally important to recognize that the way in which individuals and groups use space within the constraints of the politico-economic system is to maintain social interaction and to sustain their economic activities, including environmental resources like forests as in our case. In order to understand issues relating to Africa's tropical timber trade and the forest as the resource base, policy makers and key actors—national and international—need to understand the range of social processes which operate at contrasting but interlocking geographical scales to influence the organization and use of space (Bell, 1986, p. 16). It is in this respect that the geographic perspective adopted in the book becomes critical. This is because it seeks to sensitize the key actors in the process of decision-making of their role as being of critical importance with respect to the implications of the strategies they define, the bureaucracy they create and sustain, and the problems associated with the policies they implement. The geographic perspective also reminds them not only of the historical processes that explain the development problems prior to the twenty-first century, stressing the influence of capitalist penetration, but also the broader context within which they operate, namely the international economic system including the constraints to their actions such as those imposed by foreign trade, international aid, or loans and the implications of such constraints to African development.

It is in respect to these that the chapter discussed the importance of structural transformation in Africa, aimed at effecting spatial integration, a goal that is tied to the view that at the core of underdevelopment in SSA is the lack of fundamental sectoral linkages, which are necessary for spatially integrated development in the region (Darkoh, 1971; Mabogunje, 1981; OAU, 1982; Sawyerr, 1990). In this regard, it was pointed out that Africa's underdevelopment is a result of the colonial experience, which re-ordered the economic structures within the region's colonies to integrate them into the world capitalist system, whereby the dynamic sectors of each economy were aligned with the interests of the metropolitan countries (Sawyerr, 1990, p. 10). A lesson here is to remember the United Nations Economic Commission for Africa (UN/ECA) observation that "the structure of the African economy defines the essential features of Africa's central problem of underdevelopment" (UN/ECA, 1986, p. 1). Consequently, African leadership and key policy makers need to emphasize real structural transformation *via* spatial integration as basic to three development processes in Africa in:

- changing the region's economies from their essentially colonial characteristics,
- forging the crucial "wholeness" that characterizes advanced economies, and
- making them intrinsically self-centered although by no means autarchic.

It must be recognized that the development of spatially integrated economies would effectively undermine the accommodation of international capital in the region as was expressed as a desired goal by both the Organization of African Unity (OAU) and the United Nations Economic Commission for Africa (see OAU, 1982; UN/ECA, 1986). Indeed, it was to engender such transformation that African heads of state and of government prescribed the Lagos Plan of Action (LPA) in 1980, emphasizing "the creation of an industrial structure at the national levels within the framework of an integrated economy" (OAU, 1982, p. 21), as part of the Plan's long-term objectives—significantly by the year 2000, but evidently unmet.

Chapter 2, Turf Establishment and the Development Process: Structural Foundations of Ghana's Timber Industry, presented the historical foundations for our discussion, focusing on the development of Ghana's Timber Industry, in terms of the evolution of its structure and organization, and the role it plays in the Ghanaian economy and the nation's development process. Further, it discussed the colonial process and the associated creation of a capitalist space or turf based on the exploitation of timber and the beginnings of the nature of Ghana's tropical timber trade internally and externally. In this regard, it concentrated on the pre-independence developments involving the changes that took place in the pre-colonial and the colonial eras, ending in 1957, when Ghana secured its independence from the British, and discussed the pertinent colonial policies regarding the timber industry. For students and practitioners of African development, particularly key actors and policy-makers, situating the process of capitalist turf creation and the associated structural patterns in a geographical and historical perspective is critical since geographic space—conceptualized and treated as an aggregate of interrelated absolute, relative, and relational spaces—and development processes cannot, and therefore should not be divorced from each other, for neither can exist independently of the other (Sayer, R. A., 1985, p. 5; Gore, 1984, p. 181). This is why Gore (1984, p. 144) points out that "spatial patterns in developing countries must be analyzed historically and (should) take account of the dependent position of those countries in the international economic system," as had been suggested earlier by Slater (1974, 1975), Dos Santos (1969), and Santos (1979). This is because of the contemporary importance of the historical process in explaining the contemporary problems of African development and underdevelopment, particularly the recognition of the influence of the capitalist penetration on the African economy and society, as noted by Bell (1986, p. 16), as a basis for dealing with contemporary twenty-first century issues. It is in this respect that the chapter discussed the two related issues namely:

- that the expansion of Western capitalism was due in part to exploitation of African territories as capitalist turfs, and
- that the foreign penetration into Africa resulted in distorted, dependent economies and disrupted the traditional social system as reflected in the enhancement or creation of native authorities' coercive powers over their communities in order to enable foreign corporate access to landed resources, and to develop through conscription, the infrastructure, such as roads, railways, and ports, which are necessary for Ghana's export dependent economy (see Bell 1986, p. 17).

The discussion regarding the colonial process and the associated creation of a capitalist space or turf with respect to the exploitation of timber, and the initial nature of Ghana's internal and external tropical timber trade, serves as a springboard for reexamining Chinese resource access strategies in Africa. As was noted, China has identified "Africa as a strategic economic space"—a turf—and that part of China's geo-economic strategy is to ensure access to critical natural resources to maintain its economic growth (Carmody and Owusu, 2007, p. 506). Given Africa's colonial experience with resource exploitation, it is critical that the African leadership, policy-makers, and key actors avoid the mistakes previously made by re-evaluating the Chinese strategy so that it would not be realized to be a Trojan horse, only after the devastating conquering cargo has already been unleashed, and a very sophisticated form of neo-colonialism has been effectively established. Such re-evaluation applies to related current policies such as the US African Growth and Opportunity Act (AGOA) pertaining to American trade and investment, and new international policies that would be adopted in the future. The key lesson here is that a repetition of a new set of structural patterns, which could be problematic for national development, need to be avoided by entering into partnerships based on political and economic convenience.

Chapter 3, Independence, Instability, and Contested Turfs: Structural Changes in the Timber Industry, built on chapter 2 by discussing the key developments in the timber industry in the post-independence period through the beginning of the twenty-first century. The discussion focused on the post-independence socialist policies of the Kwame Nkrumah administration to the structural patterns that had emerged by 1983, which necessitated the adoption of neo-liberal policies at the behest of the World Bank and the International Monetary Fund, by the democratic Jerry Rawlings administration, which had morphed from a military dictatorship. This was done with respect to how these policies affected the timber trade and the associated internal and external turfs. An important aspect of this chapter is that it provides a detailed presentation of the state of the industry in terms of its problematic structural and organizational characteristics by 1983. This made it a priority target under Ghana's structural adjustment program dubbed "Economic Recovery Program" (ERP), as a foundation for the discussions in subsequent chapters relative to the issue of structural transformation. The chapter highlighted how the colonial process re-ordered the political and economic structures with the creation of an international capitalist turf with the timber industry, as one of the dynamic sectors of the economy, but had been aligned with the interests of the dominant classes of the metropolitan countries. It also drew our attention to the fact that the political and social conditions necessary to sustain and reproduce the relationships established under the colonial system had also been firmly established. Different politico-economic attempts were made by various post-independence governments like the Nkrumah and the first part of the Rawlings regimes to undo or adjust the patterns, which had become largely problematic for an authentic national development. However, an important lesson that could be learned here is the stubborn resilience of capitalism. Based on the established political and social conditions necessary to sustain and reproduce the asymmetrical relationship under the colonial system, it eventually forcefully re-surfaced in the form of structural adjustment programs—a free-market-based economic policy—in the neo-liberal era under the second Rawlings regime as the Economic Recovery Program (ERP).

A key lesson here is that attempts may be made to undo or adjust the problematic patterns resulting from the dynamics of capitalist turf creation and control, but without a carefully thought out and planned policy response or action, there is a strong tendency on the part of African governments to reproduce and sustain the international capitalist turfs by strengthening the fundamentals of their extractive economies, as did the post-independence Ghana's governments. This is exemplified by Ghana's timber industry which constitutes a major part of its economy, being opened up particularly for the transnational corporate sector, as the state ensured cheap access to the forest communities' assets on their behalf under the ERP. As was noted, this was a set of responses that resulted in the abuse of the forest community's rights, more so as the international commodity markets continued to undervalue the nation's natural resources and transfer the wealth thereby generated to international capital. Imports of food that could easily be produced by the community to meet their own needs and those of the urban communities are an example of the associated domestic economic disruption (see FERN, 2006).

Chapter 4, Turf Reclamation Strategies: neo-Liberalism and the Timber Industry, discussed the nature of neo-liberalism as an economic paradigm and tied it as the framework for Ghana's externally mandated structural adjustment program—the ERP—as a way of addressing not only the country's structural problems but as those of African economies in general. Specifically, it focused on the political economy of the ERP and the formal wood-processing industry. In order to do that, the competing perspectives on structural adjustment relative to the Ghanaian economy were presented. In connection with that, the Ghanaian political economy was situated in a global context to reiterate the argument previously made, that the problematic patterns of the industry were largely a legacy of Ghana's colonial history. The political and economic developments that culminated in the adoption of the ERP were discussed by integrating it with the locus of the wood-processing industry in the process. Based on these, a central argument was made to the effect that the ERP was essentially metonymical in that, at the time of its adoption, it constituted a convenient strategy to defuse two major problems that confronted the PNDC government by 1983, namely:

- the domestic political and economic crisis; and
- the definite collapse of the PNDC government, if it failed to accommodate international capital, as evidenced by the fall of two previous governments that had adopted a confrontational approach in dealing with international capital.

It was in respect to these that the liberal and concessionary nature of the program's macro-economic policy measures, in terms of the strategic combinations of demand management and structural policies, aimed at mitigating pre-ERP problems and fashioning a conducive environment business-wise for industry and other sectors of the economy including wood-processing were presented. A key lesson relating to this is that there is the need on the part of African leadership, as well as key actors and policy makers, to be acutely aware of the formidable pressure to succumb to or accommodate international capital, especially when the latter's control of a lucrative turf is under threat. This is more so when the country and its government are operating under dire politico-economic conditions, for which the government could also lose its own political turf. Secondly, there is the need to realize that a confrontational approach or attempts to delink might even make things worse; hence a careful, deliberate consideration of the pros and cons of the different perspectives on development policies is crucial before their adoption and implementation, especially in dealing with international capital.

Chapter 5, Turf Recapture and Consolidation for the Timber Trade: Ghana's Economic Recovery Program, discussed the adoption and implementation of Ghana's adjustment program—the ERP—and outlined and evaluated the neoliberal policy measures relating specifically to the timber industry. This was done by showing that by 1983, international capital had virtually lost the Ghanaian economy as a turf, and the PNDC government, after its attempt to delink by refusing to accommodate international capital by its socialist stance, was also losing its legitimacy and consequently, control on its political turf due to the deteriorating state of the Ghanaian economy. In an attempt to recapture their respective turfs, a marriage of convenience between these two strange bedfellows therefore developed. For the PNDC government, the only viable option was the adoption of a structural adjustment program as a strategy to resuscitate the economy. Under the program, the nature and economic contribution of the forestry sector provided a satisfying condition for it to be exploited, as a matter of mutual immediate convenience for the government and international capital. This was because of the sector's potentially rapid income-generating nature—a proverbial goose that could quickly lay golden eggs for the benefit of both parties. In fact, because of the "present gain" aspect of the Faustian bargain by the PNDC government with international capital, the goose actually laid a lot of golden eggs: the partnership led to the dramatic increase in the production and export of processed wood as was noted.

It is clearly evident that adjustment played a major role in the dramatic increase in the production and export of processed wood. These dramatic changes reflected the PNDC government's commitment to accommodating international capital, as a measure for its own self-preservation. The government was thus compelled by the Faustian bargain to have a vested interest in expanded exports. Consequently it created an enabling environment for production, provided the necessary support in respect of that, and generated pressure on the timber and wood-processing firms to achieve them. For the two strange bedfellows who were seeking to reclaim their respective turfs, processing wood for export was a convenient mutual choice in this case, because it was based on an easily accessible natural resource, and was dominated by lumber, which is essentially a raw material for further industrial processing overseas. The impact of the commitment was more evident given that the annual growth rate of manufacturing as a whole in the Ghanaian economy, mostly consisting in consumer products, fell systematically from 12.9% in 1984 to 1.5% by 1994 (ISSER, 1995, p. 98). Two key lessons can be distilled here, namely,

- the importance of a real and dedicated commitment on the part of African governments to the promotion of economic growth and development in their respective countries; and
- the entrapment of African governments into securing huge debts through foreign loans and their associated loss of control over their own economies—one of the resulting costs of the Faustian deal.

Regarding the first lesson, the dramatic increase in the production and export of processed wood demonstrates that production and export levels in various sectors of comparable African economies also have the potential for dramatic increases. Such general increases would be largely possible if African governments could really focus their energies, in the twenty-first century, on the development of such sectors, operate them with a strong commitment, and provide an enabling environment and support, as was the case in Ghana under the PNDC and NDC governments, which had a vested interest in the success of the wood processing sector. Indeed, African economies, as noted by Adjibolosoo (1993, p. 142) cannot function effectively without being upheld by a network of committed persons, who strongly believe in and continually affirm the development ideals of their society, even within the framework of controversial economic programs like SAPs. It is in this respect that Ezeala-Harrison (1995, pp. 4–5) notes, from an HF perspective, the need for African leaders to be "resolved to honest and disciplined leadership, loyalty to their countries and people, totally committed to the development of their countries . . . and must eschew selfishness and greed."

The second lesson is that in accommodating international capital, the PNDC government had access to more loans, while international capital also reaped huge profits via increased and assured interest payments on, or consistent servicing of such loans. This was apart from dictating policies including the establishment and membership of institutions such as the Timber Export Development Board (TEDB) and the Multilateral Investment Guarantee Agency (MIGA) respectively, necessary to ensure that its needs are met to safeguard its effective control over the turf. As shown in chapter 6, the adjustment program and the contributions of the forestry sector became a convenient means of the government for accessing international loans by actively servicing the country's debts with foreign exchange generated through the export drive. In 1986, three years after the adoption of the ERP, the share of official debt (i.e., government and government guaranteed debt) in total long-term debt was 85%. This debt had climbed to 93% (World Bank, 1993) in 1992, indicating increasing indebtedness as result of easier access to international loans. In accommodating international capital via the program, the PNDC government was thus able to stay in power by defusing the political crisis it faced prior to adjustment. It was even able to transform itself from a military government into a "democratically elected" one—the National Democratic Congress (NDC). Furthermore, it enjoyed "political stability" with support from the West, and became a protégé of the sponsors, making Ghana a "star pupil of adjustment" in the eyes of the World Bank and the IMF.

Chapter 6, Structural Adjustment, the Timber Trade, and Turf Degradation: Political Convenience and Desperate Deforestation, discussed the link between the neo-liberal policies, the timber trade, and deforestation, identifying the important role played by debt particularly from international loans. As was noted, Ghana's total external debt rose from US $1,398 million in 1980 to US $5,874 million in 1995 (World Bank, 1997, p. 246). Debt service as a percentage of exports of goods and services rose from 13.1% in 1980 to 23.1% in 1995, and multilateral external debt as a percentage of external debt, also rose from 19.9% in 1980 to 50.8% in 1995. These figures mean that there was a lot of pressure on the government, as part of the costs of its Faustian bargain with international capital, to meet its debt obligation, which consequently led to the excessive sacrifice of Ghana's forest as a matter of convenience. Thus the goose—the forest—was squeezed to deliver the golden eggs, the result of which was stark environmental damage, a "desperate deforestation," which environmentally seriously jeopardized our common future.

Other benefits for international capital in recapturing its turf as part of the PNDC government's Faustian bargain with international capital included the convenient contribution of the forestry sector to the generation of cheap raw materials—tropical wood and wood products—in abundance, for international capital in their home industries. In addition to that contribution, Ghana's wood processing sector served as a convenient market for products like machinery and equipment from some of the donor countries, for which the administration of the SECALs became a convenient way of sustaining, and in fact reinforcing, the market for international capital. As was noted, this market expansion and control were particularly successful with the system of tied-purchases and credit-line system (Hancock, 1989). In addition to selling products like sawmilling and other wood processing equipment to firms in the sector, some of which were inappropriate for conditions in Ghana, spare parts had to be subsequently obtained from the same sources for their maintenance. In certain cases, technicians from the seller-companies had to be brought in at the expense of the buyers to either install or repair the machinery or equipment. In quite a number of cases, the equipment purchased broke down quite frequently.

Inter alia, a major lesson for African development that must be learned from Ghana's experience for the twenty-first century is that African governments should always remember that they are mere trustees or stewards, particularly with regard to the environment; hence in formulating or implementing development policies, a prime consideration should be that they have a responsibility to both the living and those yet to be born. Consequently, it is imperative that they safeguard the environment notwithstanding the political and economic cost.

Another lesson for the twenty-first century is that the tied-purchase and credit-line systems tend to propagate the international "junk-yard syndrome," whereby inappropriate and substandard machinery and equipment or technology are dumped in developing countries for profit (Owusu 1994). Thus, while such policies make Africa a market of convenience for international capital, they make it and a dumping ground for products from the donor countries as well. It is thus in this respect that Sawyerr (1990, 28) warns that "multilateral financial institutions, as agents of 'multinational imperialism,' work to ensure that the economies of the Third World are conditioned to 'accommodate' international capital, i.e., serve the basic interests of Western industrialized economies."

Chapter 7, Turf Dynamics and Conflicts, the Timber Trade, and Spatial Integration, examined structural transformation resulting from the neo-liberal policies and how those policies addressed the goal of spatial integration. In relation to that the new trends and associated patterns of the timber trade in terms of sourcing, volume, content, and direction, and the associated turfs and dynamics that emerged as a result of the program were discussed. These new trends however were shown to have had largely negative effects not only on the environment as we saw in chapter 6, but also on the traditional interdependent relationships between the formal and informal sectors of the wood industry and on the very livelihood of the forest communities and forest management. Attempts by the government to control the traditional turfs of the forest communities and the associated informal sector was shown to be reflected in the government's effort to ban the activities of chainsaw operators, and the constant turf conflicts that now characterize the timber industry. In examining spatial integration as a form of structural transformation, the chapter documented that several changes occurred in Ghana's formal wood processing sector under the ERP with respect to linkages at both the intranational and international levels. From the perspective of the supporters of international capital, the dramatic increase in the production and exports of processed wood demonstrated "efficiency and growth," a change which was touted as evidence of the "success" of the program in Ghana. The associated changes in the spatial structure of economic activity, including changes in linkage patterns, reflected progress towards the optimum spatial production and activity pattern coincident with SAP-based economic efficiency from the perspective of international capital.

From a structuralist perspective, however, the changes do not reflect the two indicators of spatially integrated development. First, the changes were devoid of dynamic linkages within the formal wood processing industry, and with other sectors in the domestic formal economy (see Darkoh, 1973; Mabogunje, 1981; OAU, 1982). Even the new and intensive formal-informal sectoral linkage within the wood processing sector, generally absent in the economies of advanced industrialized countries, was largely exploitative, and hardly qualified as a substitute for a traditional forward linkage between primary processing and processing downstream within a formal production network. This is especially so because as a production network of any industry takes shape over time, jobs are created as linkages are established intra- and inter-sectorally—jobs that could benefit some members of the forest community to reduce high rates of unemployment, even as potential Ghanaian jobs are exported overseas via an expanded forward linkage with downstream processing firms abroad (see Fik, 1999, pp. 20–21), even discounting the potential of an integration of the informal sector with the formal.

Despite the view of international capital and its supporters that the informal activities could generate capital for self-transformation into formal operations (e.g., World Bank, 1989, pp. 135–36), so far there has been no demonstrable evidence that such transformation occurred, given their very precarious nature (Ninsin, 1991). Second, a significant decrease in the "colonial relationship" indicating a shift from external over-dependence and asymmetrical linkage relations with international capital (OAU, 1985; UN/ECA, 1986; Sawyerr, 1990) has not been apparent. So far, a clear outcome of Ghana's SAP was the dramatic growth of output in the timber sector in particular, and the cocoa and gold mining sectors—which had been Ghana's three traditional sources of commodity export earnings. Thus dramatic growth in these sectors really do not fashion real development as expected in a developing economy because the true beneficiary of the changes is international capital (Owusu, 1998a). This outcome is exactly the kind of situation that *The Brandt Report* (Brandt 1980, p. 23) referred to as it cautioned that "*one must avoid the persistent confusion of growth with development*" (emphasis mine).

The so-called "success" claimed by the sponsors of Ghana's SAP was, thus, based on a misconception (see Bromley, 1995; Owusu, 1998a), or even a misrepresentation of what development in Sub-Saharan Africa should involve. Interdependence among firms in the advanced Western economies generates various benefits (e.g., cumulative upward causation and its associated agglomeration economies), which are largely internalized because such developed economies (or blocs like the European Union) are intrinsically self-centered. In contrast, the dramatic increase in Ghana's lumber export for further processing off-shore was achieved at the cost of disrupting the basic intra-sectoral linkages between primary and secondary stages of the formal wood processing industry. That crucial linkage, at the intra-national level, was by-passed to provide huge quantities of lumber cheaply to international capital by exploiting the informal sector while the liberalized economy discouraged formal inter-sectoral linkages. At the international level, the wood-processing firms even operated as subcontractors of convenience for the "principal" firms overseas via a strong forward linkage established between them under the program, thereby consolidating the industry's asymmetrical alignment to the interests of metropolitan/donor countries.

Clearly, the SAP-based changes in Ghana's formal wood processing industry did not promote the development of the critical structures for self-sustaining, self-reliant national economic development as was expected under the LPA. In fact, Ghana's situation is not significantly different from other processed tropical wood exporting countries in SSA for at least two reasons:

- First, at the national level, the other SSA economies have structural characteristics that are similar to Ghana's in the pre-adjustment era (UN/ECA, 1986). Also, like Ghana, political power in most countries is largely vested in an alliance of social classes with no fundamental interest in the transformation of the local political economy and a break from domination by metropolitan interests (Sawyerr, 1990, p. 10).
- Second, at the international level, the other SSA economies occupy the same peripheral and subordinate position in the global capitalist economic system as Ghana, with a one-sided external economic orientation, and the associated external dependence (UN/ECA 1986; Sawyerr, 1990).

As pointed out by Owusu (2001b), Ghana's experience, based on its formal wood processing industry, suggests that the neo-liberalism-based ERP stifled spatial integration at the intranational and international levels, and sustained dependence on foreign capital and markets. An important lesson here is that the IMF, the World Bank, and the lender/donor countries or agencies are simply not charitable international institutions neither are they neutral in their activities. Their own interests as international financiers, or their need for cheap and reliable sources of industrial raw materials for their home countries, necessarily conflict with the emergence of structures for self-sustaining, self-reliant national economic development in peripheral countries. Consequently, African development within the broader global economy should be designed primarily by the Africans themselves as reflected in the LPA and the African Alternative Framework to Structural Adjustment Programs for Socio-Economic Recovery and Transformation (AAF-SAP). A more pertinent development approach, to keep African economies from the persistent stagnation, or even further decline, must address their respective adjustment to the global economy and structural transformation problems simultaneously (see UN/ECA, 1986, p. ii).

Like the goals of the AAF-SAP, this approach must be "holistic in nature" so that "the macro-economic framework, the policy directions and measures, and the implementation strategies take into account the dynamic relationships existing among all major elements related to adjustment with transformation." In so doing, "the dichotomy between structural adjustment and long-term development is eliminated" (UN/ECA, 1986, pp. ii–iii). While the peculiar characteristics of each individual country should be taken into account, great emphasis must also be put on the full mobilization and efficient utilization of domestic resources, including government assistance to the informal sector (UN/ECA, 1986). This approach, which is more consistent with the LPA, would contribute significantly to the development of spatially integrated economies in the region, and link them to the global economy in a mutually beneficial way. Unlike the Ghanaian case, where the adjustment program was adopted as a metonym, this approach requires a genuine commitment on the part of African governments to real structural transformation of their respective economies (Owusu, 2001b). Chapter 7 further showed that with the remarkable expansion in the volume of timber exported as a result of the implementation of ERP also came some diversification in terms of content, and new and diversified markets outside of Europe. However, beyond the desperate deforestation associated with the quest:

- to sustain and increase Ghana's share of the international trade in timber for the mutual benefit of the government and international capital; and
- which also created a serious domestic lumber shortage and pushed the informal sector to step in to meet domestic demand, and constricted the traditional local forward linkage between sawmills and downstream processors, the expanded volume of timber exported further marginalized Ghana's forest communities, thus jeopardizing their livelihoods.

One cannot examine Ghana's expanded international timber trade in the neoliberal era, in terms of policy implications, without due regard to the material source of the trade and its socio-economic framework, namely the forest, its peoples, and their livelihoods. In this respect, it can be said that at the root of the post-ERP trends in Ghana's international timber trade is the issue of the centralization of forest-sector governance, which is largely tainted by three interrelated and self-sustaining cycles of corruption involving the national level actors—the key decision-makers, the politicians, and technocrats—the "Timbermen," and the District Forestry Officers.

Ghana's centralized forest sector governance is a reflection of the turf conflict between the government and the forest communities in the sense that it is indicative of the basic unaddressed disconnect between the formal sector-based policy-makers, on the one hand, and poor forest communities and others, who straddle the formal and informal sectors for purposes of survival. Because of this centralized approach (despite the decentralization law, concretely expressed in the 1993 Local Government Act), members of the forest communities generally do not have a meaningful stake in the process of environmental degradation. For example, as a result of the hostility between forest community's farmers and formal sector logging, some of the former's farming practices actually tend to be inimical to the environment and the industry. As noted by Boni (2006, p. 2), as a consequence of the exclusion of farmers from timber revenue for example, "precious timber that could have been logged and exported" is felled "by farmers and simply left to rot, or more simply killed on the spot by setting fire to the base of the tree," with immense financial waste. This is because the cocoa farmers who wanted to expand their farms in the forest were of the view that the forest was an impediment to the expansion of cocoa farming, especially since the trees, they believed, belonged to the chiefs and the government, and were to be sold to timber firms.

A key lesson from this is that it is very important for the actors in policy formulation and implementation to remember that for centuries prior to the colonial era, these stakeholders have depended upon forest resources for their livelihoods—their economic production, social organization, religious beliefs, and identity have all depended upon this relationship. Consequently, the formal and regulated exploitation of the forests by the government, and the exclusion of communities from access to these forests via centralization are very disconcerting. The question by the elderly man, Opanin Dwomo, of the *Sagyimase* forest community, is quite instructive in this regard. The forest communities are frustrated with the formal sector—the government and the timber industry. "Why do we need permits to enter the forest reserves to fell trees for the construction of homes? And why does the government not do the same to our brothers who live by the sea?" They do not understand this. (Sarfo-Mensah, 2005, p. 16).

In a chapter contribution to the book, *Governing Africa's Forests in a Globalized World* (German et al., 2010), the author noted that in general, some progress had been made in the government's effort to address the general disconnect between diverse actors with new decentralization and forest management measures (ITTO, 2000; Sasu, 2004), and that measures were being taken to complement forest policy reforms (see Bamfo, 2003), but there is a need to ensure further progress (Owusu, 2010). In this regard, it was noted further that there had been an increased recognition by policy-makers that a healthy symbiotic relationship between the formal and informal wood sectors in the economy as a whole, as well as in forest communities, is critical—not only for the sustainability of the timber industry, but also for the associated domestic and international trade. As suggested in the contribution to the book (Owusu 2010, p. 227), in terms of policy recommendations, therefore:

- There must be a dedicated effort on the part of the government, and for that matter, governments of comparable African countries, to involve and cooperate with the forest communities to combat deforestation and forest degradation, in the mutual interest of both the actors and stakeholders. Bolstered with formal and informal environmental education, forest communities should be encouraged not only to seriously engage in tree replanting, but also to participate in forest management. Decentralization via such community participation contributes to the formalization of the informal sector.
- Forest communities must also be given some formal access to timber for home construction and other basic needs. This can be worked out through agreements reached among the traditional authorities, the district assemblies or locally elected officials, and other actors and stakeholders. This should be coupled with a review of revenue-sharing to ensure an equitable distribution of royalties and other benefits.
- Given the widening markets and increasing volumes of processed wood sales, a significant proportion of revenue should be expressly set aside to help forest communities exclusively in the form of social and economic infrastructure, and scholarships or financial aid for education for the youth. This would give them a vested interest in the proper management of forests. Efforts in this respect should be akin to the intensity of the PNDC government's unflinching commitment in mobilizing the forestry sector under the ERP to defuse the domestic political and economic crisis that it initially faced and to stave off its own imminent collapse as a consequence of its initial failure to accommodate international capital. Such rural poverty alleviation arrangements could be supported with agreements with major consuming countries in Europe and emerging ones such as China (which strategically needs African timber), the US, and the nouveaux riches ones such as Saudi Arabia, Kuwait, and the United Arab Emirates to set up special funds to help the forest communities and for forest management.
- Efforts to develop a well-integrated production network in terms of forward and backward linkages should be explored to address processing overcapacities and to provide more jobs. This can be done by providing tax and levy incentives to encourage new investments in downstream processing and improvements in wood recovery rates. This way, rural peoples should

be integrated within wood-processing activities to provide the youth, in particular, who are often unemployed or engage in informal activities, with jobs. Struggling domestically owned sawmills could also be financially supported to voluntarily leave the industry to reduce the level of processing capacity. Ghana's pursuit of a Validation of Legal Timber Program aimed at improving the monitoring of tree harvesting and capturing every single tree in a database for verification; a timber utilization contract system for social responsibility of concessionaires, together with other social responsibility arrangements, as well as its commitment to a voluntary partnership agreement with the European Union to improve its trade and production of legal timber are all laudable, and must be emulated by comparable African countries.

However, without addressing the partnership with the informal sector and the constituent forest communities in a decentralized fashion, success in these areas will surely be very difficult, if not impossible to achieve. Fostering such a healthy symbiotic relationship, *inter alia*, requires a new, pragmatic, and sympathetic look by African governments and policy-makers at the true nature and dynamics of the informal sector and the forest communities. In fact, the difficulties that are encountered by the government in banning the operations of chainsaw operators in Ghana are testament to this need. As noted in the FERN Report (2006, p. 17), state officials tend to play down incidents regarding the turf conflict between the government and the chainsaw operators, but it notes, such turf conflicts could completely undermine the state, and national integrity, if not checked. A lesson here from the report is that "further attempts at state repression will only accelerate this disintegration: precisely such abuse of communities created the crisis in the first place. Repression ultimately fails—especially in a state as weak as Ghana's." It is in respect to these that the Report recommends, *inter alia*:

- The participatory development of a program of local government and natural resources policy reform that cedes some measure of control over resources to democratic local institutions;
- Greater national investment in modernizing rural communities on the basis of local resource utilization; and
- A conscious effort to democratize rural institutions and to re-develop a national and even pan-African identity to counter the disintegrative effect of ethnic political agendas.

It is in this vein that Sasu (2004, p. 14) points out that the decentralization process should be "community driven, flexible and transparent" and "conscious efforts should be made to empower communities and guarantee their rights to equitable benefit-sharing." This way, policies formulated and implemented would be based on partnership, a necessary ingredient for the sustainability of the forest sector as a whole. In essence, the corruption cycles identified with respect to Ghana's timber industry, including the centralized forest sector governance and the associated timber trade is inextricably tied to the role of the human factor (HF) in development.

The lesson here for African leadership and key actors is that, despite the crucial importance of the positive aspects of HF, the role of HF in development continues to be a missing dimension in African development policies relating to real structural transformation, and thus need to be emphasized in development policies. This approach requires a genuine commitment, a basic but critical HF requirement, on the part of African governments to real structural transformation of their respective economies, on the one hand, and the international actors on the other. This is because through an effective and efficient simultaneous interaction between the positive HF aspects (the integrating core of, and key to development) and the related national and international institutions (the doors to development), the relevant rules and principles are generated to facilitate the development process (Owusu, 2004).

Chapter 8, Debt, Dependence, Turf Control, and Ghana's Economy, focused on the dynamics of debt and dependence relative to the timber industry and the Ghanaian economy in general; and their implications for comparable African countries. It was shown that if Ghana with its post-1983 timber industry was touted as a "model" adjusting country under the neo-liberal paradigm by international capital or the external sponsors of SAPs, then certainly, neo-liberalism in the form of orthodox adjustment did not suggest itself as the best remedy for the economic problems in SSA, as claimed by the proponents and international capital.

The lesson here, based on the evidence from Ghana's timber industry, is that, as a development policy for SSA economies, neo-liberalism in the form of SAPs is like the classical "Trojan horse." It tends to lure unsuspecting adjusting economies, through such mechanisms as SECALs, into accumulating substantial external debts. Further, the neo-liberal policies may actually provide a means for international capital to reap huge profits; secure assured and increased supplies of cheap raw materials; and establish dependable markets of convenience—basically securing and sustaining its turf—and the debt obligations of the adjusting economy constitute a vital part of international capital's control mechanism. It is a cost associated with the Faustian deal and a subtle form of neo-colonialism. This is because the evidence from Ghana lends credence to the view that adjustment, as a trap, constitutes an excellent basis for international capital for non-political, non-military manipulation of political and economic decision-making in the adjusting countries (Sawyerr 1990, p. 28). African governments always need to remember the axiom that "the debtor always becomes a servant of the lender," and that international loans generally constitute a mechanism for reproducing conditions of dependence and control.

In our discussion, we used Ghana's timber industry as an integrative and representative economic sector because it typifies the structural and organizational—*Turfs and Trade*—issues like the implications of the domestic and international timber trade for local livelihoods and sustainable forest management, associated with underdevelopment. Ghana's experience in relation to these from the colonial era through the neo-liberal era, and the key roles played not only by powerful international bodies such as the International Monetary Fund and the World Bank, but also indigenous ones as agents of economic change, show that there has not been much of a change in terms of the country's objective position as a subordinate component within the global capitalist economy. Ghana's post-independence economic development patterns really did not deviate much from those of the pre-independence colonial patterns. In fact, the Ghanaian economy continues to be even more indebted, and externally dependent and dominated. Developments over the period, including the politically and economically turbulent post-independence era, have not shown real structural transformation as a result of the domestic and international policies aimed at correcting the systemic economic and social imbalances that obstruct development in the country, underscoring the perspective that, the root of Africa's persistent problem of underdevelopment is the very structure of its political economy. In this respect, the geographic perspective used in this book has highlighted the important but often unrecognized or neglected role that the discipline of geography and its associated perspectives play in relation to examining and understanding the unequal relationship between the advanced and developing economies of the world; and how that relationship affects economic development and trade behavior of developing economies, like those of Sub-Saharan Africa.

Bibliography

Adjibolosoo, S. (1993). "The Human Factor in Development." *The Scandinavian Journal of Development Alternatives*, 12 (4), 139–149.

Adu Boahen, A. (2000). *Ghana: Evolution and Change in the Nineteenth and Twentieth Centuries,* Accra: Sankofa.

AGOA (African Growth and Opportunity Act) (2000). AGOA Legislation. Retrieved April 3, 2012, from http://www.agoa.gov/agoa_legislation/agoa_legislation.html. Washington, DC: U.S. Department of Government.

Alder, Dennis. (1987). "Effects of Credit Facilities on Timber Trade," Ghana Timber Export Development Board, Mimeo.

Allen, John L. (1997). *Student Atlas of Economic Development*. Guilford, CT: Dushkin/McGraw-Hill.

Alonso, W. (1968). "Urban and Regional Imbalances in Economic Development," in *Economic Development and Cultural Change*, #17 (I), 1–14, in Friedmann J. and Alonso, W. (1975).

Amanor, K. S. (1999). *Global Restructuring and Land Rights in Ghana. Forest Food Chains, Timber and Rural Livelihoods.* Research report 108. Uppsala: Nordiska Afrikainstitutet.

Amin, Samir. (1990). *Maldevelopment: Anatomy of a Global Failure*. London/Tokyo: United Nations/Zed.

Andrae, G. (1981). *Industry in Ghana: Production Form and Structure,* Scandinavian Institute of African Studies, Uppsala.

Arthur, Emelia and Kyeretwie Opoku. (2007). *Strengthening Voices for Better Choices: Ghana Country Assessment Report.* International Union for Conservation of Nature (*IUCN)* Gland, Switzerland. Retrieved November 7, 2011, from http://cmsdata.iucn.org/downloads/3_12_ghana_country_assessment_report__1_.pdf.

Atta-Mills, C. (1980). "Dependent Industrialization and Income Distribution in Ghana," In Rweyemamu, J.F. (ed.), *Industrialization and Income Distribution in Africa,* Dakar: CODESRIA.

Bamfo, R. K. (2003). Policy and Institutional Reform for Forestry: The Ghanaian Experience. Paper presented at the *Africa Forest Law Conference*, Youndé, Republic of Cameroon, October 13–16, 2003. Retrieved November 29, 2007, from http://lnweb18.worldbank.org/.../Ghana+-+Policy+&+Institutional+Reform+for+Forestry+-+Support+Document.pdf.

Barbier, Edward B, Joanne C. Burgess, Joshua Bishop, and Bruce Aylward. (1994). *The Economics of Tropical Timber Trade*. London: Earthscan Publication Limited.

Barker, K., Boateng, P., and Moreno, A. (1991). *Timber and Wood Products Section of Export Diversification in Ghana*. Report prepared for the U.S. Agency for International Development, Bethesda, Maryland: Development Associates Inc.

Bell, Morag. (1986). *Contemporary Africa: Development, Culture and the State*. New York: Longman Scientific & John Wiley & Sons.

Bird, Neil, Timothée Fometé and Gene Birikorang. (2006). *Case study 1: Ghana's experience in timber verification system design,* Verifor Case study. Retrieved November 7, 2011, from http://www.verifor.org/resources/case-studies/ghana.pdf.

Birikorang, G., R. Okai, K. Asenso-Okyere, S. Afrane and G. Robinson. (2001). *Ghana Wood Industry and Log Export Ban Study.* Forestry Commission report to the Ministry of Lands and Forestry. Forestry Commission, Accra, Ghana. 53 pp.

Boateng, P. (1991). "Potential for Increased Ghanaian Exports in Wood and Timber Exports" in Barker, K., Boateng, P. and Moreno, A. (1991). *Timber and Wood Products Section of Export Diversification in Ghana.* Report prepared for the U.S. Agency for International Development, Bethesda, Maryland: Development Associates Inc.

Boni, S. (2006). Ghanaian Farmers' Lukewarm Reforestation: Environmental Degradation, the Timber Option and Ambiguous Legislation. Paper presented at the *Colloque international "Les frontière de la question foncière"–At the frontier of land issues,* Montpellier, 2006. Retrieved October, 16, 2008, from http://www.mpl.ird.fr/colloque_foncier/Communications/PDF/Boni.pdf.

Botchwey, K. (1993). "Deregulating an African Economy: The Case of Ghana" in *Ghana Today*, Vol. 1, #1 (May), pp. 25–26.

Brandt, Willy. (1980). North-South: A Program for Survival., Report of the Independent Commission on International Development Issues. Cambridge, MA: The MIT Press.

Bromley, S. (1995). "Making Sense of Structural Adjustment." *Review of African Political Economy,* 65: 339–348.

Busia, K.A. (1953). *The Impact of Industrialization on West Africa.* West African Institute of Economic and Social Research, Ibadan. (March), pp. 31–37.

Butler, Tina. (2005). *Growing Pains and Growing Alliances: China, Timber and Africa—Timber hungry China moves into Africa.* Mongabay.com, April 20, 2005. Retrieved November 6, 2011, from, http://news.mongabay.com/2005/0420x-tina_butler.html.

Caldwell, J. (1969). *African Rural-Urban Migration: the Movement to Ghana Towns.* New York: Columbia University Press.

Cantor, D. (1967). *Import-Replacing Industrialization and Economic Development in Ghana.* Unpublished Ph.D Thesis, Harvard University.

Carmody, Pa'draig R., and Francis Y. Owusu. (2007). "Competing hegemons? Chinese versus American geo-economic strategies in Africa." *Political Geography,*

26 (2007) 504–524. Retrieved November 6, 2011, from http://ic.ucsc.edu/~rlipsch/AFRICOM/Carmody.pdf.

Carter III, Phillip. (2004). "U.S.-Africa Relations at the Beginning of the 21st Century." Remarks to the Africa Summit at the University of Miami, Miami, Florida. February 21, 2004.

Central Bureau of Statistics. (1985). *Industrial Statistics 1981–1983*. (November). Accra: CBS, Statistical Service Board.

Central Bureau of Statistics. (1983). *Industrial Statistics 1977–1979*. (July). Accra: CBS, Statistical Service Board.

Central Bureau of Statistics. (1983). *Quarterly Digest of Statistics*. Vol. I, No. 9. (June). Accra: CBS, Statistical Service Board.

Central Bureau of Statistics. (1969). *Statistical Handbook 1969*, 3rd. Ed. Accra: CBS.

Chege, N. (1994). Africa"s non-timber forest economy. *World Watch* 7, 19–23.

Chryssides, J. C. (1974). "Economic Problems of the Ghana Timber Industry with special reference to the Problem of Wastage." PhD Thesis submitted to the Centre of West African Studies and Faculty of Commerce and Social Science, University of Birmingham, England.

Coleman, H. G. and G. Amankwah. (1998). The Current Status and Future Challenges of the Development and Marketing of Timber Products: The Ghana Example. Paper presented at the 26th International Forestry Students Symposium (13–27 September). Takoradi: TEDB, mimeo.

Combé, Jean. (1972). *Forest Exploitation and Woodworking Industries in a Developing Country: Case Study of Ghana*. Memoir presented for the Diploma of the Institute, Geneva/FPRI, UST, Kumasi, Ghana.

Corbridge, Stuart. 1993. *Debt and Development*. Cambridge, MA: Blackwell.

Dag Einar Thorsen and Amund Lie. (Undated). *What is Neo-Liberalism?* Retrieved October 5, 2011, from http://folk.uio.no/daget/What%20is%20Neo-Liberalism%20FINAL.pdf.

Darkoh, M.B.K. (1971). *Industrial Location in a Developing Country: A Study of the Distribution of Manufacturing in Ghana in Relation to Selected Socio-Economic Characteristics*. Unpublished PhD Thesis, University of Wisconsin, Madison.

Darkoh, M.B.K. (1973), "Industrial Strategy and Rural Development in Africa with Special Reference to Ghana." *Geoforum*. 16/ 1973.

Davidson, Basil. (1994). *Modern Africa: A Social and Political History*. London: Longman.

Dicken, P. and Lloyd, P. (1990). *Location in Space: Theoretical Perspectives in Economic Geography*. (Third Edition). New York: Harper Collins.

Dicken, P. (1988). *Global Shift: Industrial Change in a Turbulent World*. London: Paul Chapman Publishing Ltd.

Dickson, Kwamina B. (1969). *A Historical Geography of Ghana*. Cambridge: Cambridge Press.

Domson, O. and Richard P. Vlosky. (2007). A Strategic Overview of the Forest Sector in Ghana. *Louisiana Forest Products Development Center Working Paper*

No. 81. Baton Rouge, LA: Louisiana State University Agricultural Center. Retrieved November, 29, 2007 from www.lfpdc.lsu.edu/publication/papers/WP%20 81%20SWOT%20Analysis%20of%20Ghana%20Forestry%20Sector.pdf.

Donkor, B. N. (2003). *Evaluation of Government Interventions in Ghana's Forest Product Trade: A Post Intervention Impact Assessment and Perceptions of Market Implications*. Louisiana State University, Baton Rouge, LA.

Donkor, Ben N. and Richard P. Vlosky. (2003). *A Review of the Forest Sector in Ghana,* Louisiana Forest Products Development Center Working Paper #61, Louisiana State University, Baton Rouge, LA. Retrieved November 7, 2011, from http://www.lfpdc.lsu.edu/publications/working_papers/wp61.pdf.

Dos Santos, T. (1969). "The Crisis of Development Theory and the Problem of Dependence in Latin America". In Bernstein, H. (ed.) (1973) *Development and Underdevelopment*. Harmondsworth: Penguin.

The Economist. (1989). *A change in Africa*. (September 3), 39–42.

Emmanuel, A. (1972). *Uneven Exchange: A Study of the Imperialism of Trade.* New York: Monthly Review Press.

Engeberg-Pederson, P., P. Gibbon, P. Raikes, and L. Udshot. (1996). *Limits of Adjustment in Africa*. Portsmouth NH: Heinemann.

Ewusi, K., (1987), *Structural Adjustment and Stabilization Policies in Developing Countries (A Case Study of Ghana's Experience in 1983–1986)*. Legon, Ghana: ISSER.

Ezeala-Harrison, Fidelis. (1995). "Human Factor Issues in the History of Economic Underdevelopment." *Review of Human Factor Studies*. Vol. 1, No. 2, pp. 1–25.

FAO (Food and Agriculture Organization). (1999). *Description and Quantification of Current Markets for Tropical Timber in Europe*, Ch. 3 (Table 3.4) in Markets for High-Value Tropical Hardwoods in Europe. Retrieved November, 4, 2011, from http://www.fao.org/DOCREP/003/Y2199E/y2199e06.htm.

FERN (Forests and European Union Resource Network). (2006). *Forest Governance in Ghana: An NGO Perspective*, Report produced for FERN by Forest Watch, Ghana. Retrieved November 29, 2007, from www.fern.org/media/ documents/ document_3643_3644.pdf.

Fieldhouse, D.K. (1986). *Black Africa 1945–1980: Economic Decolonization & Arrested Development.* London: Allen & Unwin,.

Fik, Timothy. (1999). *The Geography of Economic Development: Regional Changes, Global Challenges*. 2nd ed. New York: McGraw Hill.

Food and Agriculture Organization (FAO). (1988). *An Interim Report on the State of Forest Resources in Developing Countries*. Rome, Italy: FAO.

Forestry Commission of Ghana. (1994). Forest and Wildlife policy, Ministry of Lands and Forestry—Ghana. Retrieved November 16, 2011, from http://76.12.220.51/ library_info.php?doc=43&publication:Forest%20&%20Wildlife%20Policy&id=15

Forestry Commission of Ghana. (Undated). Cited in Univ. of Ghana.

Forestry Commission. (1988). *A Review Report on the Economic Recovery Programme: The Forestry Sector, 1982–87*. Report prepared by the Forestry Commission for the Ministry of Finance and Economic Planning. (April). Accra: Government of Ghana.

Forestry Research Programme. (2006). *Chainsaw Logging and Milling in Ghana: Background Study Report.* Retrieved November 29, 2007, from www.illegal -logging.info/uploads/FRP_Chainsaw_Logging_Ghana.pdf.

FPIB (Forest Products Inspection Bureau), (1991). Export Permit Report for the Year 1991. Takoradi: FPIB.

FPIB (Forest Products Inspection Bureau), (1990). Export Permit Report for the Year 1990., Takoradi: FPIB.

Frank, A. G. (1967). *Capitalism and Underdevelopment in Latin America*. New York: Monthly Review Press.

Frank, A. G. (1969). *Latin America: Underdevelopment or Revolution*. New York: Monthly Review Press.

Friedman, M. (1980). *Free to Choose*. New York: Harcourt Brace Jovanovich.

Friedman, M. (1962). *Capitalism and Freedom*. Chicago: University of Chicago Press.

Friedmann, J. (1972). "A General Theory of Polarized Development". In Hansen, N. (ed) (1972) *Growth centers in Regional Economic Development*. New York: The Free Press.

Friends of the Earth, (1993). *Plunder in Ghana's Rainforest for Illegal Profit*. Vols. 1 & 2. London: Friends of the Earth.

George, S. (1992). *The Debt Boomerang: How Third World Debt Harms Us All*. Boulder, CO: Westview Press.

George, S.1988. *A Fate Worse than Debt*. Hammondsworth: Penguin.

German, Laura, Alain Karsenty, and Anne-Marie Tiani (eds). (2010). *Governing Africa's Forests in a Globalized World,* for The Center for International Forestry Research (CIFOR). London, U.K: Earthscan.

GNA (Ghana News Agency). (1992). Ghana's Forests to be depleted by 2030. *People's Daily Graphic*. (March 10). No. 12842: 1& 8.

GTMO (Ghana Timber Millers Organization). (2000). "Effects of interventions in domestic log pricing on processing costs." Presentation by the President, Ghana Timber Millers' Organization at a Workshop on local pricing of logs at Wood Industry Training Centre (WITC), Kumasi—Ghana, organized by Forest Products Inspection Bureau, June 2000.

GIC (Ghana Investment Centre). (1990). *The Investor: A Journal on Investment in Ghana,* Vol. 1, #1, Accra: GIC.

Gordon, W. A. (1966). *Tenurial Consideration in Tropical Forestry*. The Evolution of Foresthold Tenures. Madrid, Sexto Congreso Forestal Mondial.

Gore, C. (1984). *Regions in Question: Space, Development Theory and Regional Policy*. London: Methuen.

Government of Ghana. (1984). *Economic Recovery Programme, 1984–86: Review of Progress in 1984 and Goals for 1985, 1986*. Report Prepared for the Second Meeting of the Consultative Group for Ghana., Paris, December, 1984. (November), Accra.

Green, R. H. (1987). *Stabilization and Adjustment Policies and Programmes. Country Study 1: Ghana*. Annankatu, Finland: WIDER.

Green, R. H. (1991). *Reduction of Absolute Poverty: A Priority Structural Adjustment*. Institute of Development Studies. Discussion Paper 287. Sussex: IDS.

Griffiths, I. L. (1995). *The African Inheritance*. London: Routledge.

Gwynne, Robert N. and Cristobal Kay. (1999). *Latin America Transformed: Globalization and Modernity*. London: Arnold.

Gyimah-Boadi, E. (1991). "State Enterprises Divestiture: Recent Ghanaian Experiences." In Rothchild, R. (ed.), (1991), *Ghana: The Political Economy of Recovery*. SAIS African Studies Library, Boulder, Colorado: Lynne Reinner Publishers.

Hamilton, C. (1989). "The Irrelevance of Economic Liberalization in the Third World." In *World Development*, Vol. 17, No. 10, pp. 1523–1530.

Hancock, G. (1989). *Lords of Poverty: the Power, Prestige, and Corruption of the International Aid Business*. New York: The Atlantic Monthly Press.

Harvey, D. (2005). *A Brief History of Neoliberalism*. Oxford: Oxford University Press.

Hayek, Friedrich A. (1973). *Law, Legislation and Liberty: A New Statement of the Liberal Principles and Political Economy. Volume I: Rules and Order*. London: Routledge.

Hayter, T. (1989). *Exploited Earth: Britain's Aid and the Environment*. London: Earthscan.

Hirschman, A.O. (1958). *The Strategy of Economic Development*. New Haven, CT: Yale University Press.

Hogg, D. (1993). *The SAP in the Forest: The Environmental and Social Impacts of Structural Adjustment Programmes in the Philippines, Ghana and Guyana*. London: Friends of the Earth.

Holt- Jensen, Arild. (1999). *Geography. History and Concepts: A Student's Guide* (Third Edition). Sage Publications: London.

Huq, M.M. (1989). *The Economy of Ghana.*, London: Macmillan Press.

Hutchful, Ebo. (1996). Ghana. In *Limits of Adjustment in Africa*, eds. Engeberg-Pederson, P., Peter Gibbon, Phil Raikes and Lars Udshot, 143–214. Portsmouth, NH: Heinemann.

Hyden, Goran. (1980). *Beyond Ujamaa in Tanzania: Underdevelopment and an Uncaptured Peasantry*. Berkeley: University of California Press.

International Monetary Fund (IMF). 2000. *Ghana: Statistical Appendix*. IMF Staff Country Report No.1. Washington, DC: IMF.

ISSER (Institute of Statistical, Social and Economic Research). (1995). *The State of the Ghanaian Economy*. Legon, Ghana: ISSER.

ITTO (International Tropical Timber Organization). (2005). *Status of Tropical Forest Management*: *Ghana*. Retrieved November 21, 2011, from http://www.fao.org/forestry/10901-0cf0b6c0c8ef9cdbb91e0ae827b0b9718.pdf.

ITTO (International Tropical Timber Organization). (2000). "Year 2000 Objective: The Progress Made," Tropical Forest Update, Vol. 10, No. 2, 2000/2. Retrieved November 7, 2011, from http://www.itto.int/files/user/tfu/back_issues_pdf/TFU.2000.02.English.pdf.

ITTO (International Tropical Timber Organization). (Undated). *ITTO Report Finds Improvements in Tropical Forest Management*. Retrieved November 7, 2011, from http://www.p2pays.org/ref/40/39745.pdf.

IUCN (International Union for Conservation of Nature). (2006). Strengthening Voices for Better Choices. Bangkok, Thailand: The World Conservation Union. Retrieved October 16, 2008, from www. http://cmsdata.iucn.org/downloads/strengtheningvoicesbrochure.pdf.

IUCN (International Union for Conservation of Nature). (1992). *The Conservation Atlas of Tropical Forests in Africa*. Singapore: Simon & Schuster.

Kapur, I., Hadjimichael, M. T., Hilbers, P., Schiff, and Szymczak, P. (1991). *Ghana Adjustment and Growth, 1983–91*. International Monetary Fund Occasional Paper 86. Washington, DC: IMF.

Kayode, M. O. (1987). "The Structural Adjustment Programme (SAP) and the Industrial Sector." in Philips, A. O. and Ndekwu, E. C. (eds.) *Structural Adjustment Programme in a Developing Economy: The Case of Nigeria*. NISER: Ibadan.

Killick, T. (1978). *Development Economics in Action: A Study of Economic Policies in Ghana.*, London: St. Martins Press.

Koffi Smith, E. (1996). "The Evolution of Policy into Practice. A Case Study of Ghana's Forest Policy: The Interim Measures to Control Illegal Timber Harvesting Outside Forest Reserves." Paper presented at the African Forestry Policy Forum, August 29th–30th, 1996, Nairobi, Kenya.

Konadu-Agyemang, K. and Panford, K. (eds). (2006). *Africa in the 21st Century: Pertinent Socio-Economic and Development Issues.* Aldershot, Hampshire UK: Ashgate International.

Legon Observer. (1972). Vol. 7. Feb. 11–24, 1972.

Lensink, Robert. (1996). *Structural Adjustment in Sub-Saharan Africa.* New York, London: Longman.

Lewis, A. W., (1967) "Unemployment in Developing Countries." In A. M. Whiteford (ed.) (1967) *A Re-Appraisal of Economic Development*, Chicago: Aldine de Gruyter Publishers Inc., pp. 1–17.

Lipton, M. (1982). "Why Poor People Stay Poor." In John Harriss (ed.), *Rural Development: Theories of Peasant Economy and Agrarian Change*. London: Hutchinson.

Mabogunje, A.L. (1981). *The Development Process: A Spatial Perspective*. Holmes and Meier: New York.

Marfo, E. (2004). "Coping with illegality: Conflicts over chainsaw milling and coping strategies." In K.S Nketiah, A. Wieman and K.O. Asubonteng (eds.). Chainsaw lumber production: a necessary evil?, Workshop proceedings 2, November 11, 2003, Tropenbos International, Wageningen, the Netherlands. Retrieved November 17, 2011, from http://www.tropenbos.org/publications/chainsaw+lumber+production:+a+necessary+evil%3F.

Meagher, K. (1995). Crisis, Informalization and the Urban Informal Sector in Sub-Saharan Africa. *Development and Change*. 23: 259–284.

Ministry of Information (1968) *Government Policy on the Promotion of Ghanaian Business Enterprises*, Accra, Ghana: Ministry of Information.

Mkandawire, T. (1988). "The Road to Crisis, Adjustment and De-industrialization: The African case." *Africa Development.* Vol. 13, #1, pp. 5–31, Dakar, Senegal: CODERSRIA.

MLF (1996a). Forestry Development Master Plan 1996–2020. Ministry of Lands and Forestry, Accra–Ghana.

MLF. (1996b). Circular on air-dry levy. Ministry of Lands and Forestry, 3 March 1996, Accra–Ghana. Cited in Donkor and Vlosky (2003).

Moser, C. O. (1978). Informal Sector Petty Commodity Production: Dualism or Dependence in Urban Development. *World Development.* 6(9/10): 1041–1064.

Myrdal, G. (1957). *Economic Theory and Under-Developed Regions*. London: Duckworth.

Mytelka, L. K. (1989). "The Unfulfilled Promise of African Industrialization." *African Studies Review*. Vol. 32, #3, pp. 77–137.

New World Encyclopaedia, Retrieved November 11, 2011, from http://www.newworldencyclopedia.org/entry/Colonialism.

Ninsin, K. A. (1989). "Ghana Under the PNDC: Delinking or Structural Adjustment?" in Azzam Mahjoub (ed.), *Adjustment or Delinking?: The African Experience*. The UN Univ./Third World Forum: Studies in African Political Economy. London: Zed Books, pp. 131–56.

Ninsin, K.A. (1991). *The Informal Sector in Ghana's Political Economy*. Accra: Freedom Publications.

Nkrumah, K. (1965). *Neo-Colonialism: the Last Stage of Imperialism*. London: Nelson.

OAU (Organization of African Unity). (1982). *Lagos Plan of Action for the Economic Development of Africa, 1980–2000*. Geneva: International Institute for Labor Studies.

OAU (Organisation of African Unity), *Africa's Priority Program for Economic Recovery 1986–1990*, 1985.

Odoom, F. K. (2005). *Chainsawing in the Natural Forests of Ghana: An Assessment of the Socio-economic Impacts of this Practice*. Case Study Report. Rome: FAO.

Ofei-Nkansah, K. (1991). "Ghana"s Timber Boom" in *West Africa Magazine*, Feb., 1991, p.229.

Opschoor, Johannes B. (1994). De Bretton Woods-instellingen en het milieu. In Preadviezen van de Koninklijke Vereniging voor de Staathuishoudkunde, *Vijfit jaar na Bretton Woods*. Utrecht: Lemma, 93–119. Cited in: Lensink, R. 1996. *Structural Adjustment in Sub-Saharan Africa*. New York; London: Longman.

Osei, William Y. (1997). Human-environmental impacts: Forest degradation and desertification. In *Geography of Sub-Saharan Africa*, ed. Aryeetey-Attoh, S., 35–60. Upper Saddle River, NJ: Prentice Hall.

Oteng-Amoako, A. A., Francis K. Odoom, Gene Birikorang, Kweku Pra Ghartey, O. K. Boateng Poku, Ghassan Bitar, Roger Cardoso, John H. S. Kwofie, and Francis Yaw Opoku. (2008). Voluntary Partnership Agreement Timber Industry Restructuring Group Final Report. Accra, Ghana, June, 2008. Retrieved November 22, 2011, from http://76.12.220.51/VPA_2/assets/file/IRWG%20Final%20DRAFT%20july2.pdf.

Ould-Mey, M. (1996). *Global Restructuring and Peripheral States: The Carrot and the Stick in Mauritania*. Lanham, MD: Rowman & Littlefield.

Owusu, J. Henry. (2010). "Trends in Ghana's International Timber Trade: Some Implications for Local Livelihoods and Sustainable Forest Management." Chapter 13 (pp. 261–81) in German, Laura, Alain Karsenty, and Anne-Marie Tiani, eds. *Governing Africa's Forests in a Globalized World*, for The Center for International Forestry Research (CIFOR). London, U.K.: Earthscan.

Owusu, J. Henry. (2006). "Political/Economic Convenience and Unprecedented Deforestation in Sub-Saharan Africa: Development Lessons from Ghana's Adjustment Program." Chapter 10 (pp. 161–81) in Konadu-Agyemang, K. and Panford, K. eds. *Africa in the 21st Century: Pertinent Socio-Economic and Development Issues*. Aldershot, Hampshire UK: Ashgate International.

Owusu, J. Henry. (2004). "Structural Adjustment sans Spatial Integration: Another Sub-Saharan African Development Policy Debacle." Chapter 4 (pp. 75–110) in Adjibolosoo, S. B-S. K. (ed.), *The International Development Program of Activities: What are We Doing Wrong?* Bloomington, IN: 1st Books.

Owusu, J. Henry. (2001a). "Economic Recovery, Debt and Dependence in African Economies: The Case of Ghana," Chapter 2 (pp. 17–39), in Columbus, F. (ed.), *Politics and Economics of Africa, Vol. I*, Huntington, NY: Nova Science Publishers.

Owusu, J. Henry. (2001b). "Determinants of Export-Oriented Industrial Output in Ghana: The Case of Formal Wood Processing in an Era of Economic Recovery." *The Journal of Modern African Studies.* Vol. 39, No. 1, pp. 51–80.

Owusu, J. Henry. (2001c). "Spatial Integration, Adjustment and Structural Transformation in Sub-Saharan Africa: Some Linkage Pattern Changes in Ghana." *The Professional Geographer.* Vol. 53, No. 2, pp. 230–247.

Owusu, J. Henry. (1998a). "The Myth of Development Under Ghana's Structural Adjustment Program: Evidence From The Formal Wood-Processing Sector (1983–1991)," Chapter 12 (pp. 193–216), in Adjibolosoo, S. B-S. K. and B. Ofori-Amoah (eds.), *Addressing Misconceptions About Africa's Development: Seeing Beyond The Veil.* Lewiston, NY: The Edwin Mellen Press.

Owusu, J. Henry. (1998b). Current Convenience; Desperate Deforestation: Ghana's Adjustment Program and the Forestry Sector. *The Professional Geographer,* Vol. 50, No. 4, pp. 418–436.

Owusu, J. Henry. (1995). "Structural Adjustment and African Development: Evidence from Ghana's Formal Wood-Processing Sector." *Review of Human Factor Studies*. Vol. 1, No. 2, pp. 60–79.

Owusu, J. Henry. (1994). *Ghana's Economic Recovery Program and Changes in the Structural and Organizational Patterns of the Formal Wood-Processing Industry (1983–1991)*, (PhD Dissertation), University of Iowa. Ann Arbor, Michigan: University Microfilms International.

Poore, D. and T. H. Chiew. (2000). *Review of progress towards the Year 2000 Objective*. Twenty-eighth Session of the International Tropical Timber Council, 24–30th May, Lima, Peru.

Potter, Robert B., Tony Binns, Jennifer A. Elliot, and David Smith. (1999). *Geographies of Development.* Singapore: Addison Wesley Longman.

Quarcoo, P. K. (1990). "Structural adjustment Programmes in Sub-Saharan Africa: Evolution of Approaches." In African Development Review, Vol.2 No. 2 (December).

Quartey, Jonathan D. (2010). "*Chainsaw Milling in Ghana: Assessing its Economic Impact.*" European Tropical Forest Research Network (ETFRN) News 52. Ch.4.6, 159–165. Retrieved November 11, 2011, from http://www.etfrn.org/etfrn/newsletter/news52/Chapters/4.6-Quartey.pdf.

Rado, E.R. (1985). *Ghana Revisited.* Centre for Development Studies, Univ. of Glasgow.

Reed, David., ed. 1992. *Structural Adjustment and the Environment*. Boulder, CO: Westview Press.

Richardson, H. W. and P. M. Townroe. (1986). "Regional Policies in Developing Countries" in P. Nijkamp (ed.) *Handbook of Regional and Urban Economics*. New York: North Holland, Elsevier Science Publishing Co.

Rodney, W. (1972). *How Europe Underdeveloped Africa*. Harrare: Zimbabwe Publishing House.

Roemer, M. (1981). "Dependence and Industrialization Strategies." *World Development*. Vol. 9, #5, pp. 429–34.

Rothbard, M. ([1962/1970] 2004). *Man, Economy, and State: A Treatise on Economic Principles—Power and Market: Government and the Economy*. [Two books republished in one file.] Auburn, Alabama: The Ludwig von Mises Institute. Retrieved October 5, 2011, from http://www.mises.org/rothbard/mespm.PDF.

Ryan, P. A. (2006). *Notes on High Forest Management: Mid-Term Review Report*, Ghana National Resources Programme, Accra.

Santos, M. (1979). *The Shared Space*. London: Methuen.

Sarfo-Mensah, P. (2005). *Exportation of Timber in Ghana: The Menace of Illegal Logging Operations*. Milano, Italy: *Fondozione Eni Enrico Mattei*. Operations. Nota Di Lavoro 29. Foundations Eni Enrico Mattei. Retrieved November, 29 2007, from www.feem.it/NR/rdonlyres/7B28B8FB-0D28-4FFC-9527-B966B2A54C02/1462/2905.pdf.

Sasu, O. (2004). Decentralization of Federal Systems in Forests and National Forestry Programme: The Case of Ghana. Paper presented at the Interlaken Workshop on Decentralization in Forestry, Interlaken, Switzerland, 27-31 April, 2004. Retrieved October 16, 2008, from http://www.cifor.cgiar.org/publications/pdf_files/interlaken/Oppon_Sasu.pdf.

Sawyerr, A. (1990). *The Political Dimension of Structural Adjustment Programmes in Sub-Saharan Africa*. Accra: Ghana Universities Press.

Sayer, A. (1986). "New Developments in Manufacturing: The Just-in-Time System," *Capital and Class* 30, pp. 43–72.

Sayer, R.A. (1985). "Industry and Space: A Sympathetic Critique of Radical Research." *Environment and Planning D: Society and Space*. Vol. 3, pp. 3–29.

Schwarz, G. (1990). "Stabilization Policies Under Political Transition: Reform Versus Adjustment in Brazil 1985–89." *The Developing Economies*. XXXVIII 1 (March).

SECO (State Secretariat for Economic Affairs). (2004). "Sustainable trade of tropical timber: Poverty reduction and conservation of tropical forests: Why and how SECO intervenes?" Retrieved November 6, 2011, from http://www.tropicalforests.ch/files/news/SECO_2006_booklet_en.pdf.

Seers, D. (1969). "The Meaning of Development." *International Development Review*. No. 4 (December).

Seers, D. (1977). "The New Meaning of Development." *International Development Review*, 19, 3: 2–7.

Siedman, A. (1971). *Industrial Strategy and Rural Development*. Occasional Paper, Land Tenure Center. Madison, Wisconsin.

Siedman, A. (1974). "The Distorted Growth of Import Substitution Industry: The Zambian Case." In *Journal of Modern African Studies*, 12, pp. 601–31.

Slater, D. (1974). "Contribution to a Critique of Development Geography." *Canadian Journal of African Studies*., #8 (2) 323–54.

Slater, D. (1975). "Underdevelopment and Spatial Inequality: Approaches to the Problem of Regional Planning in the Third World." *Progress in Planning*. #4 (2).

Statistical Service. (1991). *Quarterly Digest of Statistics*. Vol. IX, No. 2 (June). Accra: Statistical Service.

Statistical Service. (1989a). *Industrial Statistics 1985–1986*. (December). Accra: Statistical Service.

Statistical Service. (1989b). *Quarterly Digest of Statistics*. Vol. VII, No. 2 (June). Accra: Statistical Service.

Steel, W.F. and J.W. Evans. (1984). *Industrialization in Sub-Saharan Africa: Strategies and Performance.* World Bank Technical Paper #25. Washington, DC: The World Bank.

Steel, W.F. (1988), "Recent Policy Reform and Industrial Adjustment in Zambia and Ghana." *The Journal of Modern African Studies,* 26 (1988), 157–64.

Storper, M., and Richard Walker. (1992). *The Capitalist Imperative: Territory, Technology, and Industrial Growth.* Cambridge, MA: Blackwell Publishers.

Stutz, Frederick P., and Barney Warf. (2007). *The World Economy: Resources, Location, Trade and Development* (5th Edition). Upper Saddle River, NJ: Pearson Prentice Hall.

Szerezewski, R. (1965). *Structural Changes in the Economy of Ghana, 1891–1911.* London: Allen & Unwin.

Szerezewski, R. (1966). "The Macro-Economic Structure," in *A Study of Contemporary Ghana. Vol. I (The Economy of Ghana)*, ed. by W. Birmingham et al. London: Allen & Unwin.

Tandon, Y. (1985). "Some Theoretical Aspects of the Delinking Debate and the Experience of Zimbabwe and Uganda in this Context" Mimeo.

TBI-Ghana. (2003). *Chainsaw Lumber Production: A Necessary Evil.* Wageningen, the Netherlands: Tropenbos International. Retrieved November 29, 2007, from www.tropenbos.nl/publications/FactSheets/ChainsawInfosheetstuur.pdf.

TED (Trade and Environment Database of the American University) Case No. 193: *Ghana Forest Loss.* Retrieved November 16, 2011, from http://www1.american.edu/TED/ghana.htm.

TEDB. (1990). *The Ghana Timber Industry: Basic Information, Facts and Figures.* Takoradi: TEDB.

TEDB. (1991). *Ghana Hardwoods: Trade and Industry Directory.* London: TEDB.

TEDB. (Undated). Ghana: *Forests, Wood and People*. London: TEDB.

Tropenbos International (2004) Workshop Proceedings 2 "Chainsaw Lumber Production, A Necessary Evil?" 11 November 2003 (K.S. Nketiah, A.Wieman, and K.O. Asubonteng, Tropenbos international, Wageninen, The Netherlands. Retrieved July 9, 2012, http://www.tropenbos.org/publications/chainsaw+lumber+production:+a+necessary+evil%3F

UN/ECA (United Nations/Economic Commission for Africa). 1986. *African Alternative Framework to Structural Adjustment Programmes for Socio-Economic Recovery and Transformation (AAF-SAP).* E/ECA/CM.15/6/Rev.3., New York.

UNIDO. (1986). *Industrial Development Review Series: Ghana. Regional and Country Study Branch*, UNIDO, PPD.18, (Dec. 29), New York: United Nations.

UNIDO. (1991). *Investor's Guide to Ghana*, Report Prepared in Cooperation with the Government of Ghana., Vienna: UNIDO.

United Nations. (1970). *Economic Bulletin of Africa*. 10, 1. New York.

University of Ghana, (Undated). "REDD in Ghana: An Independent Monitoring Report by Friends of the Earth-Ghana." Retrieved November 7, 2011, from http://www.ug.edu.gh/fos/vbrp/climate/REDD_in_Ghana.pdf.

Wallerstein, I. (1974). *The Modern World-System*. New York: Academic Press.

Warren, B. (1973). "Imperialism and Capitalist Industrialization." *New Left Review*. #81.

WCED (World Commission on Environment and Development). 1987. *Our Common Future*. Oxford: Oxford University Press.

Weiss, L. 1987. Explaining the Underground Economy, State and Social Structure. *Journal of British Sociology*, 32(2): 216–234.

Wolpe, H. (1975). "The Theory of Internal Colonialism," in Oxaal, J., Barnett, T., and Booth, D. (eds.) *Beyond the Sociology of Development*. London: Routledge & Kegan Paul, 229–52.

World Bank. (1981). *Accelerated Development in Sub-Saharan Africa.*, Washington, DC.

World Bank. (1984). *Ghana: Policies and Program for Adjustment.* Washington, DC: World Bank.

World Bank. (1985). *Ghana: Towards Structural Adjustment, Vol. II: Statistical Appendix.* (Oct. 7, 1985) Report No. 5854–GH. Washington, DC: World Bank.

World Bank. (1988). *Staff Appraisal Report, Ghana—Forestry Resource Management Project.* World Bank Western Africa Department, Washington, DC.

World Bank. (1993). *World Debt Tables: External Finance for Developing Countries, 1 and 2*. Washington, DC: World Bank.

World Bank. (1997). *World Development Report 1997*. New York: Oxford University Press/ The World Bank.

World Bank. (1989). *Sub-Saharan Africa: From Crisis to Sustainable Growth. A Long Term Perspective Study*. Washington, DC: World Bank.

World Bank. (1991). *Ghana: Progress on Adjustment,* (April 16, 1991) Report No. 9475-GH. Washington, DC: World Bank.

World Resources Institute (WRI). 1994. *World Resources: 1994–95*. New York: Oxford Univ. Press.

Yeboah, Ian E. A. (1998). What's Informal about the Informal Sector? Culture and the Construction of Development in Sub-Saharan Africa. *East African Geographical Review*. 20 (1): 1–10.

Younger, S. D. (1989). "Economic Recovery Program—A Case Study of Stabilization and Structural Adjustment in Sub-Saharan Africa," in *Successful Development in Africa—A Case Studies of Projects, Programs and Policies.* Economic Development Institute of the World Bank: Analytical Case Studies. #1.

Index

About the Author

J. Henry Owusu is development geographer and associate professor in the Geography Department, University of Northern Iowa, Cedar Falls, Iowa. He has a PhD in geography from the University of Iowa, Iowa City, Iowa. He earned his BA (economics and geography) from University of Science and Technology in Ghana, and an MS (geography and regional planning) from Indiana University of Pennsylvania. He studied public policy evaluation and analysis at the Graduate School of Public and International Affairs, University of Pittsburgh; and urban studies at the School of Urban and Public Affairs, Portland State University, Oregon. He also has an MA in geography (regional development) from the University of Iowa. Dr. Owusu previously lectured in industrial geography and regional development at the Faculty of Social Sciences, Kwame Nkrumah University of Science and Technology, Kumasi, Ghana; taught human geography, and planning and geography of underdevelopment at his alma mater, the University of Iowa, and also taught geography of economic development, global issues in human geography, and geography of Africa at the University of Connecticut, Storrs, Connecticut. His research focuses on industrialization, international development policies and development in Sub-Saharan Africa, particularly Ghana; tropical timber trade; the environment and development in Sub-Saharan Africa. His other interests include regional economic restructuring; and decentralization, local livelihoods, and sustainable forest management in Sub-Saharan Africa. He has publications in several scholarly journals, such as *The Professional Geographer*, *Journal of African Economies*, *The Journal of Modern African Studies*, *African Geographical Review*, and *Review of Human Factor Studies*. He has also contributed peer-reviewed chapters to several books on African development, such as *Addressing Misconceptions about Africa's Development: Seeing Beyond the Veil*; *Governing Africa's Forests in a Globalized*

World; *Africa's Development in the Twenty-First Century: Pertinent Socio-Economic and Development Issues*; *Politics and Economics of Africa, Vol. I*; *and Water Resource Conflicts and International Security*. Dr. Owusu is a member of the Association of American Geographers and a Research Associate of the International Institute for Human Factor Development.

Lightning Source UK Ltd.
Milton Keynes UK
UKOW05n0209190813

215500UK00006B/26/P

9 780739 174012